BACKGROUND TO BEEKEEPING

This is probably the clearest textbook on beekeeping that has ever been written. Although primarily for the beginner and for those who have not yet started to keep bees, there is no doubt that the more experienced beekeeper will find much to interest him for Dr Waine writes about his hobby with great authority and a quiet humour.

All the equipment used and the operations performed in the first two years of beekeeping are described in careful detail. Having kept bees for over twenty years and tested various methods Dr Waine has evolved a basic technique which, if followed by the beginner, should guarantee success. 'Beekeeping', the author tells us, 'provides in abundance all that a hobby should give . . .' and although it is possible to make a certain amount of money from one's hobby even in a small way, as he points out ' . . . the major pleasures of beekeeping cannot be seen in a profit and loss account.'

BACKGROUND TO
BEEKEEPING

ALLAN C. WAINE
Ph.D., B.Sc.

ILLUSTRATED BY
DAVID HUTTER

BBNO
Bee Books New & Old
Burrowbridge, Bridgwater

*First published in 1955
by W. H. & L. Collingridge Ltd
2-10 Tavistock Street London WC2
and in the United States of America
by Transatlantic Arts Incorporated
Forest Hills, New York
Text printed in Great Britain by
Robert MacLehose & Co Ltd
University Press Glasgow
and illustrations by
Lowe and Brydone Ltd
London NW10*

*This edition published
by Bee Books New & Old
in 1975
Reprinted 1979
Revised edition 1987
Text printed by
Butler & Tanner Ltd
Frome and London*

© *Allan C. Waine, 1975 and 1987*

ISBN 0 905652 18 5

CONTENTS

	Foreword: The author's aim	9
I.	The sweets we seek	11
II.	The private life of the honey bee	23
III.	The public life of the honey bee	33
IV.	The keeper of the bees	40
V.	The tools of the trade	46
VI.	Starting with bees	55
VII.	Plan of campaign	62
VIII.	Swarms	73
IX.	Special manipulations	82
X.	The harvest	91
XI.	Troubles	100
	Glossary	111
	Index	115

ILLUSTRATIONS

1. Bee taking nectar from a flower   *facing page* 16
2. Digestive system of a bee   16
3. Queen, worker and drone bees   16
4. Queen, worker and drone cells   16
5. Development of eggs and grubs   17
6. Comparative times of development of queen, worker and drone   17
7. Section across a 10-frame brood chamber in summer   17
8. A typical swarm   17
9. The sting of a bee   32
10. Bee space   32
11. Protective veils   32
12. Dimensions of British standard frames   32
13. Types of foundation and methods of mounting   33
14. Methods of mounting frames   48
15. The W.B.C. hive   49
16. The National hive   49
17. The Smith hive   64
18. Stands for single-walled hives   64
19. Restricting the entrances to hives   64
20. Crown boards   65
21. Types of queen excluder   65
22. Types of smoker, and a hive tool   65
23. Travelling box for bees   80
24. Hiving a swarm   80
25. Estimating the weight of a hive   80
26. Examining a double brood chamber for queen cells   81
27. Turning over a frame   81
28. Knocking bees off a frame   96

29.	'Rapid' feeder	*facing page* 96
30.	Clearer board and bee escape	96
31.	The Pagden method of dealing with swarm and parent stock	97
32.	The Demaree method of swarm control	100
33.	A postal queen cage	101
34.	The queen cage in position	101
35.	Uncapping knife	101
36.	Honey extractors	101
37.	Strainer and bottling tank	108
38.	Uncapping	108
39.	American foul brood	109
40.	European foul brood	109
41.	Acarine disease	109

FOREWORD

THE AUTHOR'S AIM

This is a book for the beginner and for those who have not yet started to keep bees. It gives a picture of the strange world in which the honey bee lives, of its private life and of the flowers it visits. It outlines the responsibilities, duties and pleasures of a beekeeper, not overlooking the ordeal of being stung. It tells you what equipment you will need, how much it will cost and how to acquire bees.

One of the attractions of beekeeping is that its methods have never become standardized and nearly every beekeeper has his own system of keeping bees; this flexibility leads to confusion in the beginner's mind because he receives so much conflicting advice. This book sets out in careful detail the operations which a beginner will require during his first two year's beekeeping. By following the suggested basic methods, he should meet with success and should, after two years, be in a position to decide for himself the relative merits of other available methods.

Beekeeping, like many other specialized activities, has acquired a vocabulary of its own and this sometimes makes beekeeping books difficult for beginners to understand. Wherever a word or phrase with a specialized beekeeping meaning occurs for the first time in this book it is printed in italics and its meaning is explained in the text at that point. After that the word or phrase is regarded as normal but, in case you have forgotten its meaning, or in case you do not read the book consecutively, these words and phrases are gathered together in a glossary at the end.

My thanks are due to Messrs E. H. Taylor of Welwyn, Herts who supplied much of the apparatus from which the illustrations were prepared, to Mountain Grey Apiaries Ltd, of South Cave, Brough, Yorks, for the loan of an illustration of the Smith hive, and to Mr C. C. Tonsley, Editor of the *British Bee Journal*, who provided other material for illustration.

CHAPTER ONE

THE SWEETS WE SEEK

SUGAR IN NATURE

Every green leaf throughout the length and breadth of the land is one of Nature's factories for the manufacture of sugar. A leaf can absorb carbon dioxide from the air; it receives a supply of water from the roots of its parent plant and, by an alchemy never yet imitated in the laboratory, it can combine the two together to give sugar. Yet, in spite of this prodigal scale of production, sugar has always been difficult to come by and for many centuries honey was the only means of sweetening readily available to mankind.

Most of the sugar produced is combined with itself to give cellulose which forms the wood and fibrous part of the structure of a plant; such sugar is lost as a human food.

A smaller proportion of sugar is turned into starch and stored up by the plant as a reserve food for the next generation. The ears of corn, the tubers of the potato, roots and nuts are the storehouses of the plant's starch. Sugar in the form of starch is one of the essential human foods but its sweet characteristics have been lost.

A still smaller proportion of the sugar produced circulates in the sap of the plant and, in a few cases such as the sugar cane, sugar beet and maple, sugar may be won by collecting and concentrating such sap. In the fruit-bearing trees sugar accumulates in the fruit.

A minute proportion of the sugar produced by some plants—usually flowering plants—is secreted with the apparent object of attracting pollinating insects. This secretion is known as *nectar* and this is the source of the sugar which we beekeepers seek through our chosen instrument, the honey bee.

THE SECRETION OF NECTAR

In spite of all that the poets say and the hymn-writers sing, bees do not gather honey from flowers. They gather nectar from flowers (Figure 1), carry it back to the hive and there convert it into honey.

The plant organ which secretes nectar is known as the nectary. Often it is green in colour and in the buttercup may be seen at the base of the petal though it is situated at different points in different flowers. Often it lies deep down in the flowers so that an insect has to push its way right inside to reach it. In tube-shaped flowers such as the clover and honeysuckle nectar accumulates and is often visible. In America the tulip tree secretes nectar so freely that it may be removed from the flower by hand with a teaspoon; unfortunately this tree does not yield nectar in Britain but even here the lime tree, mainstay of the urban beekeeper, will sometimes yield nectar so freely that it drips upon the ground and forms a visible damp circle around the base of the tree.

The sugar content of nectar varies enormously. It differs with different species of plants and with varieties in the same species. It is affected by weather, by the nature of the local soil, by altitude and aspect.

The palate of the bee is surprisingly insensitive to sweetness and it does not seem to appreciate a nectar containing less than 7–10% sugar; drink a cup of tea containing 10% of sugar (that is about 20 lumps) and you will see how superior to the bee we are in this respect. Pear and plum nectars contain about 20% sugar but these are soon abandoned by the bee when apple nectar with 40% sugar becomes available while the dandelion (50% sugar) sometimes tempts the bee from the apple.

When the temperature is low, nectar secretion diminishes. When the weather is wet nectar may be diluted or washed out of those flowers which provide no protection against rain. In cold and wet weather moreover bees are reluctant to venture out on foraging expeditions and little nectar comes into the hive. A fine summer generally means a good honey harvest. Warm nights and slightly overcast days are probably the best conditions of all.

SOME NECTAR YIELDING PLANTS

When a plant or group of plants is yielding nectar abundantly and when the bees are gathering it eagerly the bee-keeper says that a *nectar flow* is on; sometimes he will talk less accurately of a *honey-flow* but we know already that bees gather nectar not honey. If the beekeeper knows what plant is yielding nectar he will refer more succinctly to the 'clover flow' or the 'lime flow'. A good beekeeper will know which are the major honey plants in his own district, the approximate date at which their flows occur and he will so manage his bees that he has plenty of foragers ready for the occasion.

Stores of honey are accumulated in the hive only during the four months April–July. If the season is a late one some nectar may be brought in during August while, where bees have access to heather, honey will continue to be stored until mid-September. The harvest period then is brief. It opens with the fruit blossom and (neglecting the heather) it closes with the lime, blackberry and willow herb. Nectar-yielding plants such as the willows have been in flower before the fruit blossom but any nectar so obtained is used by the bees for their immediate purposes. Plants still continue to yield a little nectar after July but, since the beekeeper (the heather-men excepted) takes off his honey early in August, the bees are allowed to add the resulting honey to their winter stores.

Here is a list of the major nectar-yielding plants in Britain, set out in their order of flowering:

Fruit blossom   Lime trees
Dandelion   Willow herb
Hawthorn   Blackberry
Sycamore   Red clover
White clover   Heather (Ling)

Fruit blossoms of the most importance are plum, pear, cherry and apple. When the earliest blossom comes out and the bees begin to work it freely the beekeeper heaves a sigh of relief since, with the new season's nectar coming in, the need for depending upon winter stores is over and the danger of starvation is past. With the advent of the new nectar the queen's rate of egg laying is greatly increased and the importance of the fruit-blossom nectar is that it enables a colony to build up its strength, make good the winter mortality and to fit itself for the work of the coming season.

In fruit growing districts the beekeeper may get surplus honey from the cherry and the apple since these yield the richest nectar.

There are so many varieties of fruit trees that the total flowering period may extend from late March to mid-May. A given variety, however, rarely yields nectar for more than 14 days and an individual flower will not yield nectar for more than a day or two.

Dandelion is a noxious weed to the gardener; to the beekeeper an invaluable source of early pollen and nectar. But even the beekeeper prefers to see dandelions growing on land other than his own. The dandelion begins to flower in early March, reaches its peak in April and continues sporadically throughout the summer. At night and during wet cold weather the flower closes up, thus protecting the nectar.

Hawthorn is a well-known hedging plant and wild shrub, growing readily anywhere even under difficult conditions. It follows after the fruit blossom and comes into flower sometime in May though the time of its initial flowering is rather variable. It would be far more valuable to the beekeeper were it not so fickle and it is commonly said that it yields nectar only once in five years.

Clover, in spite of its sensitivity to soil conditions and in spite of its requiring a high average day/night temperature before it will yield nectar, is the pre-eminent honey plant of Britain. It has been stated that 75% of the honey gathered in this country is clover honey. Clover yields little or no nectar when growing on an acid or clay soil though it grows freely everywhere. Districts rich in lime such as the Cotswolds, Chilterns and the Downs give the best bee pasture and it is significant that many of the largest apiaries are situated in these districts and upon their fringes. White clover is a valuable fodder plant and it is often sown in permanent pasture and leys. It usually escapes cutting and in this it differs from red clover which is usually cut just at the moment when the bees are working on it. Red clover has another disadvantage in that its flowers are longer and the hive bee's tongue cannot reach the bottom of the tube-shaped florets; only when the plant is yielding nectar so freely that the floret is half-full is red clover of use to bees.

Charlock flowers throughout the season and fields yellow with the flowers of this weed are a familiar sight. This plant may be esteemed by the beekeeper but it is loathed by the farmer and his use of selective weedkillers on cornfields has diminished the importance of charlock as a honey source. Rape, which is now being extensively cultivated, may well replace charlock as a honey source. It belongs to the same family as charlock and yields the same kind of quick-granulating honey.

Lime trees are for the town beekeeper the most important source of all. They begin to yield in early July requiring a fairly high day/night temperature and appear to secrete most nectar before midday. The major drawback to the lime tree is the fact that it secretes nectar on the surface of the leaves as well as within the flowers. Bees gather the former freely and the resulting honey is blackish-green and of an inferior flavour (see honeydew).

Blackberry grows freely everywhere and has a very long period of flowering from early July until October.

Willow Herb, also known as fireweed from its habit of springing up where heath and woodland fires have taken place, also grows on slag heaps and on derelict city sites. Willow herb spreads both by light, airborne seeds and by spreading underground roots. It has a prolonged flowering season beginning in early July and it yields nectar even when the average temperatures are low. Do not let either the beauty or the honey-yielding qualities of this plant tempt you to grow it in your garden; to so do is like trying to make a domestic pet of a panther.

Heather is a general name applied both to bell heather (erica) and to the tall spiky ling (calluna); both yield nectar but when the beekeeper refers to 'the heather' he is thinking of ling. This is a very important source of nectar in Scotland and the north of England, while smaller amounts are obtained from other moorland areas, e.g. Wales and Devonshire. Many beekeepers who live remote from such areas take their bees to the heather; this in spite of the fact that heather is very fickle and only occasionally gives a good nectar flow. Heather is in flower from early August until the end of September and since it begins to yield nectar when nearly all the other plants have ceased, heather honey is one of the few kinds that may be obtained practically pure.

Other Plants that may be important locally are sycamore, field beans, thistles, sea lavender, wild thyme, sanfoin and raspberry.

You will notice that late May and early June are not covered adequately by any major reliable honey plant. In districts where the white clover does not yield nectar this gap becomes larger and may include most of the month of June. Beekeepers use the phrase 'the June gap' to describe this period during which the bees may use up the surplus honey gathered from the fruit blossom and may even face starvation if the gap is unduly prolonged.

NECTAR AND HONEY

The bee has a complicated hollow tongue by which it sucks up nectar from the flowers it visits. This nectar is swallowed and passes down the gullet (see Figure 2) of the bee until it reaches the *honey sac*. This is merely an enlargement of the gullet which acts as a container in which nectar is carried back to the hive. The honey sac is sometimes referred to as the honey stomach but this name is misleading since no process of digestion takes place in it. At the hinder end of the honey sac is a one-way valve communicating with the stomach proper; by this the bee may let a little of the collected nectar pass into its stomach for its own consumption but under no circumstances can the stomach contents pass back into the honey sac and nectar unloaded in the hive is therefore quite uncontaminated.

Nectar is an almost colourless, thin liquid containing 20% to 50% sugar. Honey is a sweet, thick, yellow, amber or brown liquid containing 80% to 85% sugar. It is obvious that one major operation in the production of honey is the concentration of nectar so that its average sugar content is raised from 40% to 85%. To bring this about the bees will place only a small quantity of newly gathered nectar in each cell and this will be spread about and be painted on the walls of the cell with their tongues so that a large area of nectar is exposed to the warm atmosphere of the hive and the evaporating process proceeds rapidly. This is not only helped by the fact that the bees keep their hive warm (about 95° F.) but even more by the fact that a system of forced ventilation is in operation to remove the moisture laden air. Just outside the entrance to the hive, bees may be seen, standing with their heads inwards, digging their toes into the alighting board to prevent their becoming airborne and fanning vigorously with their wings.

Bees have to bring in 3–4 lb. of nectar to produce 1 lb. of honey.

While nectar is being concentrated another more subtle change is taking place. If you go to your grocer and ask for a pound of sugar you will be given a white crystalline powder, sweet tasting and soluble, which may be obtained from the beet or the cane; from either source the product is identical and it is to this, the commonest and best known of sugars, that the simple word sugar is applied. A scientist calls this *Sucrose* to distinguish it from other kinds of sugar. *Glucose* is another type of sugar that is familiar to most people. It is used as a food for invalids since the human body can assimilate it with extreme ease; it is less sweet and less soluble than ordinary sugar. *Fructose* is not a well-known sugar since it is not readily available and too expensive for general use. It is the sweetest of all sugars—three times sweeter than ordinary sugar—and so soluble that one pint of water will dissolve five pounds of it.

Sucrose is a compound sugar which can be broken down to give equal parts of the two simpler sugars glucose and fructose. Sucrose

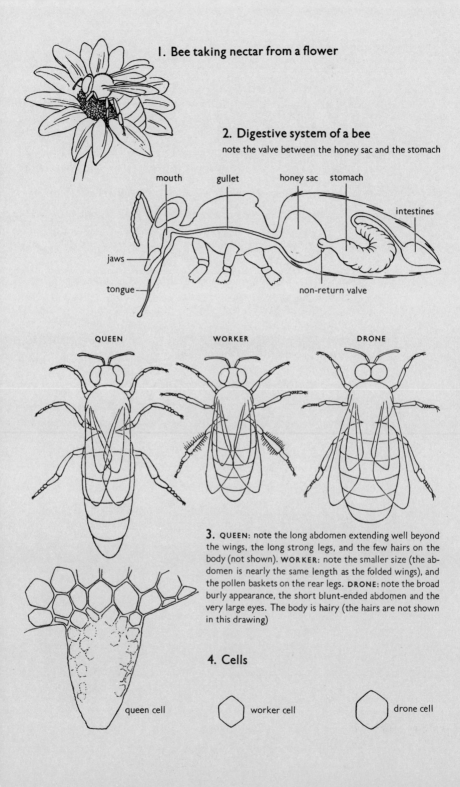

5. Development of the egg

The egg

FROM ABOVE | SIDE VIEW

 1st day: stands upright in cell

 2nd day: begins to lean over sideways

 3rd day: lying down on base of cell

 4th day: egg hatches. Tiny grub nearly invisible and lying in a pool of brood food

 5th day: grub larger. More visible but still surrounded by brood food

 6th day: grub very visible. Brood food no longer present

 7th day: grub nearly fills bottom of cell

 8th day: grub completely fills base of cell

9th day: cell sealed. Grub not visible.

6. Times of development (in days) of queen, worker and drone

egg
larva unsealed
larva sealed

7. Section across 10-frame brood chamber in summer

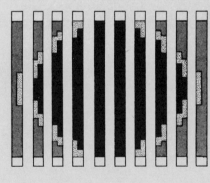

honey
pollen
brood and eggs

8. A typical swarm

is the principal sugar occurring in nectar and its breakdown is the second major operation brought about by the bees when converting nectar to honey.

While nectar is being carried from the flower to the hive, the bee adds to it a minute trace of a substance known as diastase, capable of breaking up sucrose into its two component parts, glucose and fructose.

THE CHARACTERISTICS OF HONEY

When all the changes described in the previous section are complete the bees cover over with wax the cells which contain the finished honey. Such honey is described as *ripe* or *fully ripened* while material in all stages between nectar and honey is described as *unripe*.

Ripe honey, fresh from the combs is a thick viscous liquid the colour of which may vary from nearly water-white through yellow to dark brown; green and red tints are not uncommon. This variation in colour is due to the plant which contributed predominantly to the honey, some plants, e.g. clover and willow herb, always give a pale honey, while others, e.g. apple and blackberry, give dark honey.

On keeping for a month or two, and especially when the weather grows cold, clear liquid honey goes solid and crystals appear throughout the whole mass. When this happens the honey is said to have *granulated* and it is a matter of personal taste whether liquid or granulated honey is preferred. Certainly granulated honey is less sticky and more manageable to eat; moreover, it can be piled up more thickly upon a foundation of bread and butter. Those who prefer their honey liquid or *clear* as it is usually called, can convert the granulated honey by keeping it in a warm place (not more than 120° F.) for about 24 hours. The crystals which appear are glucose, the less soluble sugar present, while the fructose remains in solution and fills up the spaces between the crystals.

Honeys vary very much in their tendency to granulate. Some, such as rape, will granulate within two or three days after extraction from the cells. Apple and blackberry honey granulate slowly and incompletely. Heather honey rarely ever granulates.

Honey was designed as a food but most of the other things that we eat—meat, roots, tubers, seeds, nuts, fruit—were produced by animals or plants for quite a different purpose but we have used them in spite of this. Bees had to produce a food which would keep good through the long winter months. Honey is sterile and yeasts, moulds and bacteria which fall into liquid honey either die or remain dormant. Honey in an effectively sealed container will keep indefinitely. There is in my possession a specimen 20 years old and older specimens must exist.

A winter food for bees must be highly concentrated, partly for economy of storage and partly because frequent meals would be impossible under winter hive conditions. One pound of honey has the food (or calorific) value of $1\frac{1}{4}$ lb. of bread, 23 eggs, or half a hundredweight of cabbage.

Since bees leave their hive during the winter to empty their bowels only at infrequent intervals (a period of two months is not uncommon) honey must contain little or no waste matter. It is almost completely assimilable, the only indigestible residue being the shells of a few pollen grains which get in accidentally.

When the weather is cold, bees are inactive though they are not actually asleep and, under these circumstances, their food must be easily digested. Honey is the most easily assimilated food known; it can be absorbed into the body without any action on the part of the stomach; it is sugar in the form the body needs it.

Finally, honey tastes delicious, though whether this is important to the bees we cannot say. All these advantages benefit us as much as they do the bees.

SOME SPECIAL HONEYS

Since the flowering periods of nectar yielding plants overlap or even coincide it is very difficult to obtain unmixed honey from one plant alone. Moreover, a beekeeper generally takes off his honey from the hive at the end of the season and honey from all sources is blended during the process of extraction. Much of the subtle and exquisite flavour of British honey is due to this natural blending.

Nevertheless, it is interesting to know the characteristics of honey from the major honey sources and under certain unusual conditions individual honeys have been isolated. Such conditions obtain where bee-forage is largely confined to one plant such as in cherry growing districts, in downlands with predominant white clover and in the heather districts. Sometimes the weather gives the necessary selective conditions by suppressing by low temperature and rain the nectar flows of plants just before and just after the plant under investigation. In all these cases the beekeeper must take off the honey collected at very frequent intervals so that later nectar flows do not become mixed with honey from the source in which he is particularly interested. As a beginner you are not advised to try these methods, but you may like to note the characteristics of honeys from individual sources so that you may be able to determine the major constituents of your own honey.

Tree Honey. It so happens that most of the early sources of honey are trees or shrubs and the combined honey from these sources is often called tree honey. It includes pear, plum, cherry, apple, hawthorn and sycamore. Such honey is rather dark in colour and granulates slowly; when granulated it looks coarse and unattractive but in the liquid state its flavour is excellent.

Hawthorn Honey is amber coloured, of exquisite flavour and its aroma fills the whole hive. A blend of apple and hawthorn honey is (this is a personal opinion) the finest honey in the world, surpassing even the clover and the heather.

Clover Honey is one of the most saleable and popular. In the liquid state it is an attractive clear yellow colour and crystallizes to a pale fine-textured mass giving the finest granulated honey of all; but its flavour is inferior to that of the tree honeys or heather honey.

Charlock or Rape Honey is pale and not very characteristic in flavour, but it blends well with the stronger flavoured honeys. It granulates so readily that it is an embarrassment to the beekeeper.

Lime Honey is medium coloured and often has a greenish tinge. It has a characteristic after-flavour reminiscent of peppermint which many people find attractive. Although this honey originates from a tree it is never included in the tree honeys.

Heather Honey is in a class by itself. It is dark amber in colour and jelly-like in texture, so much so that a jar of it may be inverted and nothing will run out, yet on stirring with a spoon it becomes liquid but settles down to its jelly-like state once again if left undisturbed. The indescribable, but quite unforgettable aroma of heather honey is at once apparent when a jar of such honey is opened. The flavour of heather honey is unique and the slightly bitter after-taste repels some people and attracts others. Like oysters, heather honey is an acquired taste. As a connoisseur's delicacy heather honey commands a price 25% to 50% higher than that of any other honey.

Honeydew. If your honey is dark green or black, thick and tastes like treacle then it is almost certainly contaminated with *honeydew*. Honeydew is the name given by beekeepers to honey produced from sweet secretions which, although not true floral nectar, are gathered by bees. There are many small insects which feed on the sap of plants and trees and excrete sweet juice which is a slightly modified sap of the plant; this when gathered by the bees and treated as true nectar gives rise to honeydew. It has a poor flavour but may be used for cooking. It is not a good winter food for bees since it contains too much matter which they cannot digest.

Plant secretions from the surfaces of leaves, from the joint of the leaf stem and twig and from the tips of the leaves of conifers also give rise to honeydew and this, although dark in colour, may be of good quality; indeed, countries with large pine forests depend upon these trees for a very considerable proportion of their honey crop. Honey from this source is also generally known as honeydew in Britain.

Undesirable Honeys. Privet gives a dark, bitter honey which will spoil other honey with which it may be mixed. Fortunately, privet is usually kept so closely cropped that it seldom has a chance to flower. Ragwort gives a deep yellow honey with a nauseating taste. *Rhododendron ponticum* gives a poisonous honey in some parts of the world but, in spite of its profusion, it appears to be ignored by bees in Britain.

BEE GARDENS

Beekeepers often ask what they can plant in their garden to provide additional forage for their bees. You can do very little indeed in this direction. A bee will forage within a radius of two miles of its own hive, i.e. over an area of at least twelve square miles, and your garden can be only a very small part of this total.

Nevertheless it is extremely interesting to set out in part of your garden plants that are highly attractive to bees; you will not increase your honey yield to an appreciable extent, but you will provide yourself with opportunities of watching minutely the bee at work. A bee garden is an unusual and attractive feature of any garden and your friends who vie with one another in their rockeries, ponds, herb gardens and herbaceous borders will envy your originality in setting one up. Here are a few plants that should be included in any bee garden.

Arabis (*Rock Cress*). The single variety is a valuable early source of nectar and pollen.
Aubrieta. Valuable early source of nectar and pollen.
Borage. 2 ft. high. Pale blue flowers resembling those of the potato. Bees never leave this plant when it is in flower. The leaves taste of cucumber and may be used in salad. If you like a drink known as Pimm's No. 1, a fresh sprig of borage is an essential component and this may add to its appeal.
Catmint. Bees delight in this. If grown too freely it may even flavour the honey with its own peculiar tang.
Crocus. Eagerly sought by bees for its abundant supply of early pollen.
Ivy. Flowers in October and is the latest of all nectar-yielding plants. Let it grow freely without clipping.
Limnanthes. A bright yellow and white buttercup-like flower. Best sown in autumn for early spring flowering.
Mignonette. Too well known for description.
Phacelia. A charming annual, easily grown from seed. Not very well known and all your friends will ask you what it is; this is another very good reason for growing it.
Thyme. If this is grown in the interstices of crazy paving it is difficult to avoid treading bees to death, so thickly do they cluster on it in its flowering period.
Wallflower. The single varieties are valuable to bees in the early spring.

The following shrubs are also very useful:
Cotoneaster horizontalis. Bees show great enthusiasm for the small pink flowers. Queen wasps may often be caught on this shrub.
Gooseberry. Let us call this a shrub for the sake of our list. The small green flowers give abundant supplies of early nectar.
Lavender. Actually a major source of honey in some parts of the world.
Raspberry. Gives abundant nectar during the 'June gap'.
Snowberry (*Symphoricarpos*). Worth consideration as a hedging plant but it spreads very easily and must be kept under control.
Veronica. All varieties attract bees (especially *Veronica spicata*). A long flowering period is an added attraction.

Bees tend to ignore the more highly cultivated of flowers and will often ignore double varieties while working single varieties, e.g.

arabis, clarkia and wallflower. Your bee garden will not be a blaze of colour but will be very fragrant.

Suggestions for further reading

Howes, F. N. (1945) *Plants and Beekeeping*. Faber and Faber. O.P.
Harwood, A. F. (1947) *British Bee Plants*. Apis Club. O.P.
Carter, G. A. (1945) *Bees and Honey*. 'Bee Craft' Books.

CHAPTER TWO

THE PRIVATE LIFE OF THE HONEY BEE

THE COLONY

'A hive of bees' is a common phrase but beekeepers often use the name of the home when they really mean to refer to the inhabitants, so we hear such expressions as, 'My hive has swarmed', or, 'One of my hives gave me 60 lb. of honey'. Let us not say these things. A hive is the house in which the bees live. The inhabitants together with their combs are known as a *colony*, while bees, combs and hive are referred to collectively as a *stock*; a small, newly-formed colony is often called a *nucleus* (plural: *nuclei*).

If you look into a colony in May or June your first impression will be of thousands of bees milling around and the impression is a true one. A small colony will contain 10,000–15,000 bees, an average colony 20,000–40,000 while very strong colonies may contain as many as 60,000 bees. When you ultimately own a colony of your own remember that you are the overlord and supreme ruler of this microcosm of 40,000 or more living creatures and that the power of life and death lies in your hands.

The restless activity so apparent everywhere, is partly due to the many hive duties which the bees are carrying out and we shall look into these later in this chapter. Recent discoveries have shown that through certain formal movements, known as dances, bees can communicate with one another so some of the visible activity may be put down as conversation. We must hesitate before thinking that any of their comings and goings are aimless since all activity generates muscular heat and this is the bees' method of keeping the hive warm.

A colony contains three types of bee, illustrated in Figure 3:

The Queen—the perfect female. Under normal conditions there is one queen and one only.

The Drones—the males. There are several hundred of these.

The Workers—sometimes described as neuters though they are really females with undeveloped reproductive organs. There are 20,000–40,000 of these depending upon the size of the colony.

THE QUEEN

She is a handsome insect and half as large again as a worker. Most of her greater size is due to her long abdomen which projects well beyond the tips of her folded wings. Her body is less hairy than those of the other bees and often she is of a different and lighter colour. Her legs are long and strong and she stands a little higher than the other bees. These characteristics must be clearly remembered since they help us when we are faced with the task of finding the queen among the other 40,000 inhabitants of the hive. (See Figure 3.)

Shakespeare knew that one bee was larger than the others and called it the King but when the king was observed laying eggs the name changed to Queen, egg-laying being no function of a king. Even the name Queen is misleading since she does not rule the hive. Her job is to lay eggs and thus she is mother to all the bees in the hive. Her egg-laying performance is amazing and at the height of her season she will lay up to 1,500 eggs (about equivalent to her own weight) per day. So that she may be free from all interruption she is surrounded by a circle of young workers who feed her, clean her and attend to all her wants.

Until she is mated, she is referred to as a *virgin queen*. In her virginal days she flies around freely until at last (about seven days after emergence from the cell) she takes her mating flight uniting with the drone while both are on the wing. When they are joined together they fall to the ground where they remain for perhaps 20 minutes, then the queen breaks free and returns to the hive alone. If the queen is not satisfied with her first mating she will go out on a second mating flight but once she is satisfied and has begun to lay she never mates again. In a special sac within her body she stores the drone's seminal fluid and this contains sufficient sperms to fertilize all the hundreds of thousands of eggs which she will lay during her lifetime. When the mating flights are over she remains in the hive for the rest of her life, except when a swarm emerges; then she takes wing again to join the swarming bees. If a queen is not mated within about three weeks from the date of her emerging from the cell she appears to become incapable of being mated; a long spell of weather bad enough to prevent mating flights will sometimes bring this about. Such a queen is capable of laying eggs but since she is unmated the eggs are not fertilized. Strangely enough such eggs will hatch but they always produce drones and such a queen is known as a *drone-laying queen*. The condition also arises in a very old queen who has used up all the store of sperm which she received on mating.

The queen is a long-lived insect and may live up to six years; but after her second season (or in exceptional cases her third) her rate of egg-laying drops off and she must be replaced by a young queen.

THE DRONES

The drone has become a symbol of idleness but he has a function to perform which is, if required, to mate with the virgin queen and his reward for this is death. No doubt also his activities help to keep the hive warm, but he brings in no nectar or pollen and helps himself liberally to the workers' hard-won stores.

The drone is about the same length as the worker but much broader with a blunt ended abdomen. He is more hairy than the other bees and often darker in colour. His two magnificent compound eyes are so large that they meet on the top of his head. He has no sting. (See Figure 3.)

A hive in summer contains several hundred drones even though only one or two will ever be required for mating; nevertheless, if attempts are made to eliminate all the surplus drones, the workers become listless, unsettled and do not work well. Sometime in July or August the workers decide that the summer is too far spent to raise any more queens and that drones are no longer required. Accordingly they are starved by being denied access to the honey cells and then, too weak to fly, they are thrown out of the hive to die. Truly a drone pays dearly for his short, comfortable life. If he escapes death in ecstasy then he meets his end in ignominy and starvation.

We have seen already that a virgin queen can lay eggs which will produce drones. Investigation shows that a drone-producing egg (usually but inaccurately called a 'drone egg') is never fertilized even when laid by a mated queen. When a queen is laying an egg in a drone cell she withholds her supply of sperm and we may truly say therefore that a drone has no father; this fact is of great importance in selective bee breeding.

Drones wander freely from hive to hive and may congregate in a colony containing a virgin queen.

THE WORKERS

When poets and authors use the bee as the symbol of busyness it is the worker they have in mind. Energetic when on the wing, restless within the hive and active both by day and night it is small wonder that the worker dies of overwork rather than of old age. A worker bee is born with the capacity to do a certain amount of work rather than to live for a set period. The life of a worker is about six weeks at the height of the season but a bee born in August when the year's work is practically finished will live six months or more, surviving the winter and taking up its duties in the following spring.

Of the three types of bee the worker is the smallest and its abdomen only just projects beyond the tips of the folded wings. Its tongue is long and well developed while on its hind legs it has spiny structures known as *pollen baskets* and when these are full of pollen the worker appears to be wearing coloured trousers. The worker has rudimentary ovaries and under special circumstances can lay about a dozen eggs which, if they hatch, produce drones (see Chapter 11, Laying Worker) since workers can never mate. (See Figure 3.)

Both queen and worker possess stings, since the sting is essentially a female weapon, but the queen is very reluctant to use it except against another queen.

The duties of a worker vary with its age and the order in which these duties are undertaken is remarkably constant. Here is the life schedule of the average worker.

First six days. Attending to the queen, cleaning cells, feeding older larvae (or grubs).

Sixth to fourteenth days. Workers of this age are known as *nurse bees* because certain glands become active and they are able to secrete a special food upon which young larvae and queen larvae are fed. Early in this period the worker takes its first flight outside the hive.

Fourteenth to twenty-first days. The wax-making glands become active and the main duty of workers of this age is to produce wax.

Twenty-first to forty-second day (or until death). Bringing in nectar, pollen and water. Bees of this age are generally known as *foragers*. Once a bee has begun to forage probably it never again reverts to its earlier duties, except in such an emergency as a swarm.

The reception and ripening of nectar and the storage of pollen are duties undertaken by bees of all ages.

The periods suggested above may be lengthened or shortened according to the pressure of work in the hive but the order in which the duties are undertaken never varies.

THE COMBS

In nature, bees attach their combs to the roof and sides of the recess in which they live, but the beekeeper persuades them to build within a moveable wooden frame so that any individual comb may be removed, examined and replaced without damage. Cells are built on either side of a vertical sheet (the midrib) of wax. The cells are not opposite to one another but staggered and the bases are in the form of shallow three sided pyramids as shown in Figure 4. The hexagonal plan of the cells giving the maximum economy of space and material, the uniformity of size and surprising strength of the comb are known to everybody. Indeed the uniformity of size has been over-emphasized because there are four kinds of cell in a hive and all four may occur on the same comb.

Worker cells. These are the most usual type of cell and they are about $\frac{1}{2}$ inch deep. They are hexagonal and are built almost invariably with two parallel sides of the hexagon vertical. If we measure across these vertical sides we shall find that the total width of 5 cells is exactly one inch. There are 25 cells to the square inch.

A queen lays eggs in these cells and the resulting larvae feed, grow and finally emerge as full-grown workers. Because these cells serve as nurseries to the workers the name 'worker cells' is applied to them. Bees also use them for the receipt of nectar and for the storage of pollen and honey; when used to store honey the cells are often deepened and show a tendency to slope upward from the base.

Drone cells. These too are hexagonal but are rather larger being $\frac{3}{4}$ inch deep. When measured across the vertical sides they are found to run at four to the inch or 16 to 18 per square inch. In these cells the drones are raised and, since the drone is a larger insect than the worker it requires a larger cell for its development. In addition bees use drone cells for storing honey although, strangely enough, they rarely put pollen into them. Drone cells are usually found along the bottom and in the corner of combs in the brood chamber.

Transition cells. Since both worker and drone cells may be found in the same comb and since they are different in size there must be some irregular cells to fill up the odd spaces where the two types join. Such cells are known as 'transition cells'. They are used for the storage of pollen and honey, but never for breeding.

Queen cells. The three types of cell described may be found in any hive at any time but queen cells are specially built for the raising of new queens and are torn down when the process is complete. Unlike all other cells, queen cells are used once and once only. They differ from other cells in that they hang downwards. They are not hexagonal or even angular; often they are described as 'acorn-shaped' though the resemblance is often not very marked. Queen cells occur singly either on the face or along the bottom edges of a comb. Bees will build from 2 to 20 of them according to their future plans.

Brace comb. In the natural state bees space their combs very regularly and the distance from midrib to midrib in adjacent combs is about $1\frac{3}{8}$ inch. In a hive this spacing is controlled by the construction of the frames and generally runs at $1\frac{3}{8}$ to $1\frac{1}{2}$ inch.

Bees will often build short lengths of comb at right angles to the general direction of the main combs. Such bits of comb are known as *brace-comb* and they are a nuisance to the beekeeper since they join together the main combs and prevent their easy removal from the hive.

THE BROOD CHAMBER OR NURSERY

We can now look even more closely at the comb and see for ourselves what the contents of the cell look like.

In nature the tendency of the queen is to lay her eggs in the lower part of the combs while the bees use the upper parts of the comb for the storage of the honey. In a hive the beekeeper will often make the natural division more sharp by inserting a grid of wires or a sheet of specially perforated zinc forming a barrier through which the worker bees may pass easily but which acts as a barrier to the queen. When this is done, the lower part of the hive to which the queen is confined is known as the *brood chamber* while the upper part is known as the *super* or *supers* if several chambers are used.

Eggs. Bee *larvae* in any stage of development are known by the collective name of *brood* and if we look into the nursery or brood chamber they catch the eye first; but we must take things in their logical order and look for eggs. The egg is white and about $\frac{1}{16}$ inch long; it is cylindrical and rounded at both ends but one end is a little larger than the other so that it is shaped like a young marrow. When an egg is first laid it stands upright at the bottom of the cell with its larger end uppermost. On the second day it begins to lean over and on the third day it is lying down in one of the angles at the base of the cell. At the end of the third day it hatches. (See Figure 5.)

Larvae. The young worker larva is fed at very frequent intervals by the nurse bees and grows extremely rapidly and soon has to coil itself up to lie in the bottom of the cell. By the fifth day it has grown so large that it must lie full length in the cell and over the five-day period it has increased its weight 1,500 times; if we picture a human baby growing to the size of an elephant in five days we might get an idea of the startling rapidity of the growth. Larvae, when healthy, are a beautiful pearly white colour and while living in open cells are called *unsealed brood*. (See Figure 5.)

After the fifth day the feeding stops and the cell containing the full grown grub is sealed over. The change from grub to perfect insect is hidden from our eyes; all we know is that a perfect young bee bites its way out of its cell about twenty-one or twenty-two days after the laying of the egg. These times are different for queen, worker and drone and are shown in Figure 6.

Sealed Brood. Cells containing sealed larvae are known as *sealed brood*. The cappings of the cells have a buff or fawn colour and a dull porous appearance quite unlike the pure wax cappings used for sealed honey. The cappings on worker brood are flat or very slightly raised but the cappings of drone brood are hemispherical.

Breeding cells. Newly made comb varies in colour from white to pale primrose yellow but a larva spins a dark brown cocoon which it leaves behind in the cell. Once a cell has been used for breeding it takes on a brown colour and as the process is repeated the whole comb turns nearly black. There is no harm in this and a queen often seems to prefer to lay in a used cell. The bees whose job it is to clean out the cells see that the cell walls do not become unduly thick and they remove accumulated cocoons, so that the cell wall never exceeds 0·006 inch in thickness. The wall thickness of cells of newly built comb is about 0·004 inch.

A good queen lays her eggs systematically, missing very few cells. She begins at the centre of the comb and works outwards so that an oval or round patch of brood appears on the comb. Before she reaches the edge of the initial comb she will have begun to work on the neighbouring combs so that the complete brood nest is roughly spherical as shown in Figure 7.

The brood patch is usually surrounded by a band of cells containing pollen which is kept in this convenient position since it is important food for larvae. The corners of the frame will often contain honey ready for immediate use although the bulk of the honey is above in the super.

FOOD STORES

Honey and nectar in all stages of ripening will be found in some parts of the brood chamber and more plentifully in the super.

Nectar. Nectar is nearly water-white and so thin that if the comb containing cells of it is jerked or held horizontally nectar drops out and thousands of bee-journeys are wasted.

Honey. When the conversion is complete honey is sealed over with cappings of pure wax. Some races of bees—notably those that are dark in colour—leave a thin layer of air between the honey and wax capping and if this is done the cappings look beautifully white and attractive. Other races of bees fill the cell brim full of honey so that it is in actual contact with the capping after the cell is sealed; in this case the cappings are darker in appearance but however the cell may be sealed, the cappings are quite distinct and different from the buff coloured cappings of brood.

Pollen. We have so far said very little about pollen. It is not one of the products which a beekeeper takes for himself but it is a food of the utmost importance to the bees themselves. Bees often gather as much pollen as honey and a strong colony may use over 100 lb. in a year. Much pollen is used as soon as it is brought in but some of it is stored over the winter so that brood rearing may start early in the spring, long before fresh supplies are available. Pollen will attract the attention at once, looking red, orange or yellow as it lies in the cells; these are the commoner colours, but willow herb gives a blue pollen, poppies black and blackberry greenish-white.

Bees never fill cells full of pollen but leave about $\frac{1}{8}$ inch space at the top. When pollen has to be stored for the winter this space is filled with honey and the cell is sealed over with wax in the normal way. After sealing there is no way of telling which cells contain pollen preserved in honey and which contain honey alone, unless the comb is held up to the sun when the pollen cells appear opaque. If the honey is very dark even this test is not reliable.

Pollen is used principally in the feeding of larvae since it contains nitrogen and is therefore a body-building food. Nurse bees eat pollen to enable them to secrete their special brood food which is very rich in nitrogen. After two or three days of feeding on brood food the worker larvae are weaned and are then fed on a mixture of honey and actual pollen and it is for these reasons that bees tend to store pollen in the brood chamber where it is instantly available for the nurse bees to carry it to the thousands of hungry larvae.

PROPOLIS

If you have handled a wooden frame containing a comb during our preliminary examination you will notice that it feels sticky. Bees have the habit of collecting the sticky exudations of trees, taking the resinous gummy substance back to the hive and plastering it liberally around the hive. They dislike vibration, looseness and draughts and will therefore stick their combs together and plaster up all the cracks in the hive with a material which is known in Britain as *propolis* and in the U.S.A. by the much less classical name of *bee glue*.

The name propolis is derived from the Latin prefix pro (= in front of) and the Greek word polis (= a town) and refers to the habit of certain kinds of bee in gathering excessive amounts of this substance and building a defensive wall of it in front of the entrance to their hive. Bees vary very much in their use of propolis and some gather and use very little.

Recently a considerable demand for propolis has arisen and first-class specimens will fetch as much as £1·50 per ounce.

SCENT

You will have noticed that a clean, pleasant smell arises from the hive. The sense of smell is probably the bees' keenest sense and every hive is said to have a distinguishing odour. By this means strangers are detected and denied admission unless they are (*a*) drones, (*b*) very young bees or (*c*) carrying a load of nectar.

The smell we observe is a compound smell. One component of it is aromatic and comes from propolis, especially propolis gathered from the balsam poplar. Another component arises from the honey; in particular hawthorn and heather honey permeate a hive with their particular fragrance. The smell of the bees themselves reminds me of certain types of roses but I cannot pretend to distinguish one hive from another. If the bees are angry it seems particularly marked and if the bees are disturbed they set to work to disseminate their characteristic smell by exposing the scent gland which lies towards the tip of the abdomen and fanning with their wings to spread the scent in all directions. Guided by this scent, stray bees can find their way back to their own hive more easily.

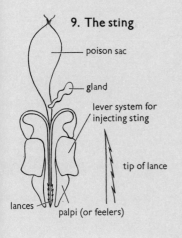

9. The sting

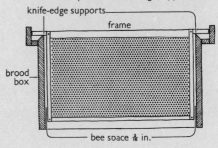

10. Bee space
section showing frame supported in brood chamber. Note bee space and knife-edge supports

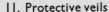

11. Protective veils

black net veil

black net veil reinforced with wire rings

wire gauze and net veil

square wire gauze or 'meat safe' veil

12. Dimensions of British standard frames

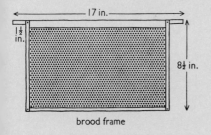

brood frame

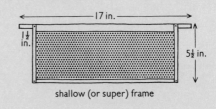

shallow (or super) frame

13. Types of foundation and methods of mounting

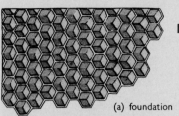

(a) foundation

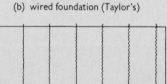

(b) wired foundation (Taylor's)

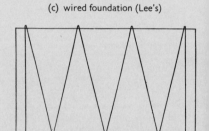

(c) wired foundation (Lee's)

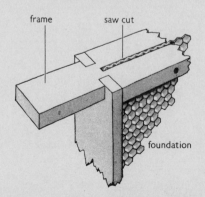

(d) foundation mounted in frame with saw cut in top bar

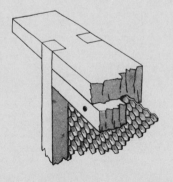

(e) foundation mounted in frame with wedge top bar

CHAPTER THREE

THE PUBLIC LIFE OF THE HONEY BEE

We must never forget that, although the details of the life of an individual bee are interesting, the outstanding thing about bees is their social life. A bee merges its individuality in the life of the colony; it acts in obedience to the dictates of the mysterious power that rules the life of the colony; it works itself to death in the service of the colony; it will willingly sacrifice its very life in defence of the colony. So, if we are to understand our bees, we must look not only at the life of an individual bee but at the history of the complete colony throughout the year.

WINTER

The important thing about the honeybee is that, unlike its cousins the bumble bee and the wasp, it does not hibernate. It remains awake and active throughout the winter months and will emerge from the hive and fly around whenever the weather is warm; even when snow is on the ground bees may be seen in flight when the sun is shining brightly and the air is warm. Nevertheless, the bee is a cold blooded creature. Its body temperature is not constant as is ours but varies with the surrounding temperature. As the air temperature falls so does that of the bee, and as its temperature falls it becomes more inactive and lethargic. If the temperature of a bee falls below 40° F. it probably never recovers and this limit must therefore be avoided at all costs.

Clustering. As winter comes on the bees conserve their natural warmth by huddling together. *Clustering* is the word that beekeepers use to describe this behaviour. When the temperature inside the hive falls below 48° F. the bees cluster. At higher temperatures the bees can be found covering the faces of the combs in the normal manner. At lower temperatures they cluster more closely.

The winter cluster consists of a spherical group of bees. This sphere is broken up into vertical slices by the combs themselves but bees always cluster in such a position that there are empty cells in the centre of the cluster and in these empty cells some of the bees remain. The bees on the outside of the cluster are in a cold position and it is not yet certain whether they change position with bees from inside or whether they die of cold and are replaced by others willing to make the supreme sacrifice.

Inside the cluster the bees are much more thinly dispersed and indeed the cluster may be regarded as hollow. The queen remains in the centre and so good is the insulation provided by the outer shell of bees that a temperature of 90° F. may be maintained inside, a temperature high enough to hatch eggs. Accordingly, the queen begins to lay a few eggs very early in February, and these hatch and develop within the winter cluster.

Since activity is reduced and hive temperature lowered, and above all, since there are no grubs to be fed, the food consumption is enormously reduced. The monthly food consumption during November, December and January does not exceed 2 lb. of honey per month and even stocks with very inadequate stores are in no danger of starvation until late February.

SPRING

When the British winter is at its worst, when blood runs cold and wise beekeepers are sitting by the fire, in early February the queen begins to lay and to raise workers for the season which instinct tells her will soon be beginning.

Cleansing Flights. Whenever the outside temperature rises to about 55° F. bees will emerge and fly around, taking the opportunity to empty their bowels; such flights are spoken of as *cleansing flights*. Bees are fanatically clean creatures and unless suffering from dysentry will not excrete within the hive. As winter relaxes its power, flights become more frequent. The winter aconite and crocus are yielding pollen while the willows, a little later, will yield both pollen and nectar. All these are eagerly visited to augment the winter stores.

Water. Another important task is the carrying of water, for it seems that bees are reluctant to eat their stored honey in concentrated form and much prefer to dilute it to the consistency of nectar before use. Around drains, ornamental ponds, puddles and ditches bees can be seen taking in supplies of water for the hive. If natural supplies of water are not available within 200 yards of the hive, a good beekeeper will provide a drinking fountain and save his bees both labour and unnecessary hardship.

Break-up of the Cluster. Early in March the winter cluster will break up and when this happens the bees have to keep the entire hive at the egg-hatching temperature of 90° F. The queen has now a much larger choice of cells in which to lay and her rate of egg-laying increases accordingly. The drain on the winter stores is now very severe. Not only are there many young larvae to be fed but the maintenance of a high temperature makes great inroads into the food reserves. It is during the month of March (and the first two weeks of April in more northerly districts or in very late seasons) that colonies die of starvation. In a natural state this would be due to an exceptionally bad season in the previous year or to the fact that the colony had remained weak and had thus not had an adequate force of foragers to accumulate the 30 to 40 lb. of honey necessary to carry it over the winter season. In an apiary it is due to the fact that the beekeeper has not replaced, by adequate feeding with sugar, the honey taken away.

First Supplies of Nectar. When the apple blossom comes out the danger of starvation is past. In the midlands it is about the second week in April but those who live in a cherry growing district or in the south may have adequate supplies of new nectar at least two weeks earlier.

The influx of the first nectar enormously stimulates the queen and in early May she reaches her maximum rate of egg laying, probably 1,500 per day.

SUMMER

Strength of Colony. It is the aim both of the beekeeper and of the bees themselves to increase the strength of the colony before the main nectar flow takes place so there is a large force of workers ready to deal with the nectar harvest. If the main nectar flow comes in mid-June the colony should be at full strength by the end of May since a bee has normally to serve three weeks' apprenticeship in the hive before it is ready to go out foraging. Moreover, three weeks elapse between the laying of the egg and the emergence of a worker and the peak laying of a queen should therefore occur six weeks before the main honey flow occurs.

Size of Brood Chamber. Under natural conditions the bees will see that the queen has ample space to lay all the eggs required by building extra combs in the lower part of the nest. If they are living in a hive the beekeeper must see that the queen is never cramped and must be prepared to increase the size of the brood chamber by adding extra combs (or a box of extra combs) if this seems necessary.

Storage Space. During the months of June and July extra boxes of combs must be added to the upper parts of the hive to provide adequate storage space for the nectar that comes in so rapidly at this time. The beekeeper's harvest is gathered during June and July and upon the weather during these two months depends the weight of the yield. It is true that beekeepers who live in a fruit district may accumulate part of their yield in April or May and beekeepers who live in a moorland district may gain additional honey during August and September but for most of us June and July are the critical months.

Swarming. As honey producers we could wish that the energies of our bees were devoted solely to gathering nectar. Unfortunately (for us) the summer season is also the season for swarming.

We have already seen that the laying power of the queen gives the bees a very effective and rapid means of increasing their number, but a means of increasing the larger unit, the colony, is also necessary otherwise accidents and the depredations of other animals such as man and bears would reduce the number of colonies until the race of the honeybee was extinct. Swarming is an act of colonization, of emigration. It is Nature's way of increasing the number of colonies in existence.

Swarming is not an annual event in all colonies. Some bees are more prone to swarm than others and beekeepers try to breed bees that are reluctant to swarm since swarming is a nuisance in a well-managed apiary.

When the bees are determined to swarm they first build a number—say 12 to 20—of queen cells and in each cell the queen lays an egg. This usually takes place in late May, June or early July. The eggs hatch, the grubs develop and are fed liberally with brood food (known as *Royal Jelly* when it is fed to queens) throughout the five days during which the larvae are unsealed. On the 9th day after the laying of the eggs the cells are sealed and at any moment after that, if the weather is suitable, the swarm may emerge.

Swarms will generally come out between 12 noon and 2 p.m. Many of the old foragers emerge, the queen joins the throng and any bee more than six days old (and therefore capable of flying) may take part. Generally about half the inhabitants of the hive join in, mainly the older bees, while the younger ones stay behind. Each bee joining the swarm fills its honey sac with honey so providing itself with about three days' iron rations.

When the bees first emerge they fly wildly and joyously about and the air is black with bees. Very soon they begin to settle until the familiar bunch of bees, clinging together and forming a tight cluster is formed. This initial cluster forms within 30 yards of the *parent colony* and remains in that position for 2 to 24 hours. During this period scouts are sent out to find a new home and, when a suitable place has been found, the whole swarm takes wing again and heads for the chosen place. (See Figure 8.)

Three points of importance for the beekeeper are:

(1) The presence of a number of queen cells in the hive is a warning that a swarm is about to emerge and the age of the cells will give the probable date of emergence.

(2) If a swarm is to be captured this must be done during the initial clustering because the new home chosen by the scouts is often outside the normal foraging range of the parent hive.

(3) It is a good idea to keep an empty hive in a corner of the apiary; if the scouts choose this for their new home the beekeeper's operations are simplified. Moreover (and this must be said in a whisper) somebody else's stray swarm may occupy the empty hive.

And what is happening in the parent hive from which the swarm emerged? Most of the foragers have joined the swarm but there are plenty of young bees who will soon become of the foraging age and in the meantime plenty of stores have been left behind and there is no danger of starvation.

All the brood, both sealed and unsealed, remains behind. From the sealed brood will soon emerge young bees to make up the depletion in population caused by loss of the swarm. There are plenty of bees of nursing age to look after the unsealed brood. Finally, there is a dozen or so queen cells from which new virgins will emerge in seven or eight days' time. The bees await the event with contentment and are not disturbed by the fact that they are temporarily without a queen.

When the first new queen emerges one of two things happens. If the bees approve of her and if the swarming impulse has exhausted itself she is allowed to go around to the other queen cells, bite a hole in the side and sting her still imprisoned sisters to death. In this work of destruction she is often helped by the workers. If the workers do not approve of the first virgin to emerge, they will restrain her murderous intentions until a second virgin has emerged. A brief, fierce fight then ensues between the rival queens in which one is killed instantly while the other walks away victorious to dispose of her remaining sisters in the usual way. This done, she settles down, takes short flights and in about seven days mates and becomes the new head of the colony.

In the second case, events take quite a different course. If the colony's urge for swarming is not satisfied they will permit the first virgin to emerge to leave the hive with a second or after-swarm, generally known in Britain as a *cast*. A cast emerges about nine days after the first or *prime swarm* and this period is accounted for by the following facts. (*a*) A prime swarm emerges as soon as the first queen cell is sealed. (*b*) Seven days elapse between the sealing of a queen cell and the emergence of the queen. (*c*) The newly emerged virgin requires a day or two to acquire her full strength. Since the prime swarm may be delayed by bad weather, the nine day interval may be shortened by a day or two.

While all this is going on the remaining virgins are prevented by the workers from leaving their cells; they protest by *piping*—uttering a shrill peep-peep-peep sound—and the free virgin joins in giving her sisters a piece of her mind. A colony may throw a second and third cast (or even more) but when its swarming impulse is finally satisfied then one of the virgins disposes of her remaining sisters and settles down to the task of heading the colony.

A cast is smaller than a prime swarm and is headed, of course, by a virgin queen. It takes longer to get established, indeed it often fails to do so and the bees die of starvation. Both swarms and casts are a nuisance in a well-conducted apiary and much of a beekeeper's time is spent in detecting, checking or dealing with swarms.

AUTUMN

The Massacre of the Drones. This is the first act of the workers' autumn programme. Nectar is coming in less freely; the queen's rate of laying has dropped to a few score eggs per day; the season for swarming has passed and with it the need to provide for the mating of a virgin queen.

Cessation of Egg-Laying. By mid-August—the date varies both with the type of bee and the season—there will be no eggs in the hive and very little brood and by September the brood chamber is empty of brood.

Preparing for Winter. Pollen gathering is still going on busily and much of it will be stored in vacant cells in the brood chamber. The bees may also move some of their stored honey down into the brood chamber. Slowly the hive temperature falls and the bees become less active until the winter cluster is formed once again.

The year's cycle is complete.

Suggestions for further reading

Maeterlinck, M. (1970) *The Life of the Bee*. Allen & Unwin.
Françon, J. (1939) *The Mind of the Bees*. Methuen. O.P.
Teale, E. W. (1943) *The Golden Throng*. Museum Press. O.P.
Frisch, K. von (1950) *Bees, their Vision, Chemical Senses and Language*. Oxford University Press.
Ribbands, R. (1953) *Behaviour and Social Life of Honey Bees*. Bee Research Association & Dover Books, U.S.A. O.P.
Butler, C. G. (1974) *The World of the Honey Bee*. Collins.

CHAPTER FOUR

THE KEEPER OF THE BEES

THE PERSONAL FACTOR

One vitally important aspect of beekeeping is too often overlooked. We may read many books of instruction on beekeeping without coming across any reference to the influence of the beekeeper himself. Yet it is certain that some people make a success of beekeeping while others, living perhaps in the same district and working with a similar strain of bee, obtain only mediocre yields of honey and ultimately give up in disgust. What then are the qualities that a beekeeper needs?

Courage is the first quality that a beginner will need; or if not courage, fortitude and stoicism, for bees have stings and the beginner must be prepared to be stung. Since the new beekeeper is not yet adept he will be stung more frequently because his mishandling will irritate the bees; even worse, the after-effects of the sting will be more prolonged and painful because the body has not yet built up the necessary resistance.

Deliberation is an essential quality. The impulsive, hasty and impatient types do not make good beekeepers. Anger and exasperation are fatal when dealing with bees. The best beekeepers are often slow spoken, slow moving, deliberate folk and to such people the handling of bees comes naturally for all movements used when manipulating a stock must be deliberate. Bees are intensely irritated by quick, jerky movements; a hand passed rapidly over the top of an open hive will often be attacked and stung when, by slow and deliberate movements, an entire comb covered with bees can be removed without any bother at all. People who back hastily away from a bee hive and who ward off inquisitive bees by striking wildly in the air with their hands are the ones who invariably get stung. Those who stand still and allow the bees to crawl over them are usually quite unharmed.

Patience you must have if you are to keep bees. You have a long apprenticeship to serve and it will be about three years before you feel competent to handle your own bees without advice from more experienced beekeepers. In the meantime you will have to cope with swarms, you may kill bees by starvation or mismanagement, you may be faced with a succession of bad seasons and in spite of all this you must persevere.

Keeping bees is not an easy way of augmenting your income and of providing your family with unlimited honey and if you imagine any such thing then please read no further for you are doomed to disappointment; but if you can face all the difficulties and disappointments that beset beekeeping and not be discouraged, read on.

ON BEING STUNG

It is the sting that keeps beekeeping select. If bees were stingless we should have every Tom, Dick and Harry keeping a hive in his garden until there were too many bees for the available nectar. If you take up beekeeping you will be regarded by your friends with a mixture of amusement and envy; amusement because the craft of beekeeping is select and beekeepers are therefore deemed a little odd; envy because you have the courage to handle bees in spite of their stings and because you have honey for your delight or for sale as the case may be. Before we can become the envy of our friends we must master this problem of stings.

The sting of the bee is in the tail (or the tip of the abdomen) as you will presently discover. The sting is essentially a female weapon and all drones are stingless. The queen possesses a sting but she is very reluctant to use it and the beekeeper may handle a queen with little danger of being stung. A bee intent on stinging emits a slightly higher and louder buzz than normal and this angry note usually serves as a warning; there is also something purposeful in its approach and a persistence in its search for vulnerable spots to attack. Quick as is the act of stinging, it is not carried out without the bee estimating the suitability of the surface by means of a pair of feelers (or palpi) that lie alongside the sting. A bee will never attempt to sting a finger nail and will generally avoid the palms of the hand where the skin is thick. (See Figure 9.)

On being stung the natural instinct is to knock away the offending bee but the sting is barbed and remains behind with the poison sac attached. It is a strange thing that, although the sting and poison sac are now detached from their original owner, the muscles go on working, burying the barbs deeper and pumping poison into the wound. This may go on for 20 minutes and it is important to remove the sting as quickly as possible. The embedded sting should be removed by scraping either with the thumb nail, the hive tool or even the nozzle of the smoker. Any attempts to pick out the sting by pinching it between finger and thumb will only compress the poison sac and inject the poison more quickly. The smell of the poison will probably provoke several other bees to sting and it is a good idea to camouflage the smell by puffing a little smoke on to the wound.

As for the bee that left its sting behind it will die, generally within half an hour. If you have sufficient self-control to refrain from brushing it away at the moment of stinging it will walk round and round, pivoted by its abdomen as though trying to extract its sting but I have never seen a bee free itself in this way, nor does it seem possible in view of the barbed shape of the sting. In about half-a-minute it tears itself free and flies away to die.

The first effect of a sting is the sharp pain produced by the penetration of the lance and the initial injection of the poison. The pain is no worse than pricking oneself deeply with a needle unless the sting is in a very sensitive place such as the eyelid or the lip. In an hour or two swelling takes place and although it may be considerable it need not cause alarm; it will die away completely in 48 to 72 hours. The attendant itching and swelling are far more troublesome than the initial sharp pain.

Fortunately, the human body is able to immunize itself against stings and you will find that with each subsequent sting the after-effects grow less marked and less prolonged until, after about a dozen stings, little is felt except the initial sharp pain; this never disappears or diminishes no matter how many times one is stung. A few unlucky people never gain this immunity, the after-effects become more serious and in addition there is a pumping of the heart, a difficulty in breathing and even unconsciousness. Medical aid is essential in bad cases and the usual remedy is the injection of adrenalin. Women seem to be more prone to become hypersensitive to bee stings than men.

Although by suitable protective clothing it is possible to make oneself bee-proof, an accidental sting is still a possibility and the risk to those who are severely afflicted by stings is so great that it is better to renounce all ideas of keeping bees.

PERSONAL PROTECTION

Veils. Like Salome the beekeeper is recognized by his veil, but if he is wise he never discards it. The face is the favourite target of the angry bee and a very vulnerable one too.

The simplest of all veils is a plain black net hanging from the brim of a hat, and long enough to tuck in under the jacket collar. The material from which a veil is made is always black because it is nearly impossible to see through a white veil in bright sunlight. The black net veil is simple and cheap and very easily removed from the hat and slipped into the pocket. Its disadvantage is that the tip of the nose tends to touch the black fabric and thus becomes exposed to attack and when the wind is blowing almost any part of the veil may touch the face, usually at the moment when a marauding bee is looking for a hopeful point of attack.

One method of overcoming the above disadvantages is to sew two rings of wire into the netting at about eyebrow and chin level so that the netting is held away from the face. Another device is to use a cylinder of wire gauze suspended from the brim of a hat but with these alternatives portability is sacrificed and personally I find these types of veil inconvenient and clumsy.

The most effective type of veil is often referred to as the 'meat safe' and consists of four more or less rectangular panels of wire gauze sewn together with tape. By means of muslin and elastic the veil fits snugly around the crown of a hat while the lower edge is extended to tuck down into the jacket or, better still, to fit over the shoulders with suitable tapes for fitting. (See Figure 11.)

A veil by itself is useless; it must be fixed to a stiff brimmed hat and a hard straw hat is the ideal support if you can get hold of one. The popular male felt hat has too flexible a brim though it may be reinforced with a ring of plywood. It is impossible to generalize about womens' headgear, but most of it is useless from the bee-keeping point of view.

Gloves. Those who are beginning beekeeping will be much reassured by wearing a pair of gloves but these must be either rubber surgical gloves (hot and sweaty) or the special gloves known as 'Birkett's'; these are made of soft white leather and attached are long white cotton gauntlets with elastic tops to make them bee-proof. Some of the more delicate operations cannot be carried out in gloves. With increasing confidence, gloves will be largely abandoned but even so it is wise to wash the hands with 'Izal' or some other carbolic disinfectant before manipulations are undertaken. A further precaution is to remove your wrist watch because bees dislike the smell of sweat and a little patch of perspiration lingers there in hot weather.

Suitable Clothing. The question of clothing is very important. Bees dislike, and show their dislike by attempting to sting, dark-coloured, rough-textured material. Skirts are impractical garments because of the bees' tendency to climb upwards; trousers plus cycle clips or breeches are much to be preferred. Perhaps the most satisfactory beekeeping garment is a white boiler suit such as is often worn by house decorators; if the buttons are replaced by a zip fastener a cool garment that is acceptable to the bees and proof against their stings is obtained.

PROFIT AND LOSS

Personally I think that the phrase 'a profitable hobby' is a contradiction in terms. A hobby should provide a change of occupation, interest, amusement and exercise and most people spend money on gardens, games or cars to obtain these ends.

Beekeeping provides in abundance all that a hobby should give, but so long as you confine yourself to two or three colonies—and a novice should not attempt more—you will be doing well if you break even. Forget all the stories you have ever heard about profits from bees paying the rent, paying the insurance or augmenting the old age pension.

As your skill increases you may increase the number of your stocks to ten or more and make a small profit. Now your hobby has become a 'side-line' and the god Mammon has pushed in his troublesome finger; the carefree days are gone. If you increase your stocks to fifty or more you will probably increase your income by an appreciable amount but your hobby has now become an 'alternative source of income' and the Income Tax Inspector will be waiting on your threshold.

To quote prices and costs invariably dates a book yet you will want some idea of outlay, running costs and returns involved in setting up one or two hives. Here are approximate figures which are true for the time of revising this book, Spring 1987.

Cost of complete hive, i.e. single brood chamber and two supers, all fitted with combs; roof and floor.

 W.B.C. £103. National £61.

Spare hive, one brood chamber, floor and roof; without combs.

 W.B.C. £84. National £48.

Miscellaneous equipment including veil, feeder, smoker, super clearer, queen excluder and hive tool. £36.

New equipment will therefore cost about £223 for a W.B.C. hive or about £145 for a National hive.

These figures cover equipment only and we still have to acquire some bees. All the suppliers of equipment listed on page 54 will also supply bees. See also pp. 58–59 for acquiring bees.

With these figures in mind we can draw out a simple balance sheet. Running costs include sugar (about 15 lb. per hive) and honey jars (£19 a gross). Also your own labour, depreciation on equipment and interest on money spent on equipment and bees.

Expenditure		*Income*	
Depreciation on equipment 10% on £200	£20·00	30 lb. honey at £1·70 per lb.	£51·00
Interest on capital equipment, £200 at 10%	£20·00	Loss	£33·63
Bees £60 at 10%	£6·00		
Labour. 12 hours at (say) £2·00	£24·00		
Sugar. 15 lb. at 25p	£3·75		
Jars. 72 at 14p	£10·08		
	£84·63		£84·63

Most beekeepers, having bought their equipment, will forget about the cost and make no allowance for interest or depreciation. Moreover, since the labour is the beekeeper's own, this item is often omitted by those who attempt to work out their running costs or the cost of producing one pound of honey.

Under these circumstances even one hive will show a profit. This is the place to repeat once again that the major pleasures of beekeeping cannot be seen in a profit and loss account.

CHAPTER FIVE

THE TOOLS OF THE TRADE

COMBS AND FRAMES

From the point of view of the bees there is only one important item of equipment and that is the combs which serve them as storage cupboards, nurseries, workrooms, and bedrooms in the winter. The shape and location of their nests seems a matter of secondary importance but the arrangement of the combs is always the same. They hang down from the roof and are supported at the sides at many points while the bottom edges of the combs are free. A beekeeper may keep bees in an upturned box, and in this they will build their combs, ultimately filling it up completely. If the box is weatherproof and adequate in size this arrangement is perfectly satisfactory to the bees but highly inconvenient to the beekeeper. A modern beekeeper wants to examine combs and some arrangement must be devised whereby they may be removed and replaced without damage. Combs may have to be removed or transferred from one hive to another and none of these operations can be carried out if the bees stick their combs firmly to the roof and sides of their home.

Part of this problem was solved by the blind Swiss naturalist Huber (1789) who induced his bees to build their combs within wooden frames each of which could be conveniently handled without damage and with very little disturbance to the bees themselves. The hive was to him now an open book and he was able to lay the foundations of our knowledge of the bees' way of life.

Following this discovery, many beekeepers tried to devise a hive into which such frames would conveniently fit, only to discover that the bees stuck them to the walls of their hives either by propolis or brace comb so that the combs once more became immovable. The difficulty of this was overcome by an American, L. L. Langstroth, (1851) who discovered that if the frame was suspended in the hive so that a gap of $\frac{1}{8}$ inch to $\frac{1}{4}$ inch was left between the frame and inside walls of the hive, the bees respected this space and did not seal it up. If the space—known as the *bee-space* and usually $\frac{3}{16}$ inch—is less than $\frac{1}{8}$ inch the bees will gum it up with propolis while, if the space is greater than $\frac{1}{4}$ inch, bees will build brace-comb and once again stop it up. Between the bottom edge of the lowest frame and the actual floor of the hive this distance may be increased to $\frac{3}{4}$ inch or more. Figure 10 shows a cross section of a modern hive with a frame surrounded by the appropriate bee-spaces and supported at only two points.

Types of Frame. The types of frame used most generally in Britain were standardized as long ago as 1880 by the British Beekeepers' Association and Figure 12 shows the two standard types. These differ only in depth and the deeper of the two ($8\frac{1}{2}$ inches deep) is commonly used in the brood chamber and is generally known as a 'brood' or sometimes as a 'deep' frame. The shallower frame ($5\frac{1}{2}$ inches deep) is generally used in supers.

The British brood frame was standardized in the days of the original British black bee, a variety which was not very prolific, and ten standard frames provided adequate breeding space for the queen. Modern bees are largely crossed with the prolific Italian types and 10 brood frames no longer provide an adequate brood chamber for the bees of today. This difficulty may be overcome by providing more than 10 brood frames in the brood chamber (see page 65) or by using larger frames.

In spite of its shortcomings the British standard frame is extremely convenient to handle and is so generally used that those starting bee-keeping would be well advised to adopt it. If you buy your first bees as a colony already established on combs you will almost certainly receive those combs built into British standard frames. While the British standard frame still remains my favourite, it is only fair to add that most of the commercial beekeepers use one of the larger sizes of frame.

Foundation. You are probably wondering how the beekeeper persuades his bees to build their combs within the wooden frames that he has so carefully provided. The secret is to fill each frame with a very thin sheet of beeswax upon which is embossed the hexagonal pattern of the base of the cells. Such a sheet of embossed beeswax is known as *foundation*. Figure 13 (*a*) shows an enlarged view of a sheet of foundation from which you can see that it is so carefully made that not only are the hexagonal outlines of the cells all perfectly reproduced but also the pyramidal base of the cell. Both sides of the foundation are embossed and when it is mounted centrally in its wooden frame, the bees build-up or 'draw-out' both sides. Foundation may be embossed with the pattern of worker cells (worker base foundation) or drone cells (drone base foundation); the latter type was once popular for use in supers, but the modern tendency is to use worker base foundation throughout the hive.

When freshly made, the foundation has the beautiful aromatic odour of beeswax and in this state the bees take to it readily and begin work on it at once. On keeping, this aroma fades away and bees are reluctant to work on stale foundation; often they will nibble away the edges or bite holes through it; sometimes they will ignore it altogether and build their combs in the spaces between the frames. Do not therefore attempt to keep a stock of foundation and do not order more than you can use up in two or three months.

Foundation is sold already cut into sheets suitable either for brood or shallow frames and the required size must be specified when ordering. It is sold by weight which gives about eight brood or fourteen shallow sheets to the pound.

Foundation is often reinforced by embedding fine wire in it; it is then better able to stand up to the handling, shaking and manipulation that the beekeeper will give to the completed combs. The manufacturer will do this efficiently and neatly for a few extra pence per pound. Beekeepers who are interested may wire their foundation for themselves, but the process is better left until skill has been acquired in other directions. Figure 13 (*b*) and 13 (*c*) show two of the most commonly used methods of wiring foundation.

Figure 13 also illustrates the two usual methods of securing foundation in frames. The most obvious method (*d*) is to put a saw cut in the top bar and to insert one edge of the foundation in this, securing it afterwards with nails or melted wax. The method shown in Figure 13 (*e*) involves an extra deep bar, part of which is cut away so that it can be removed and afterwards nailed back in position after the insertion of the foundation; while frames of this type are a little more expensive they are the most satisfactory.

When using wired foundation the bottom bar of a frame is always split so that the foundation is supported along its two longer edges. Some modern types of frame also have grooves in the side members so that the foundation is supported on all four sides; these are very good indeed.

Most beekeepers buy their frames 'in the flat'. The component pieces are all cut accurately by machine and the beekeeper has to

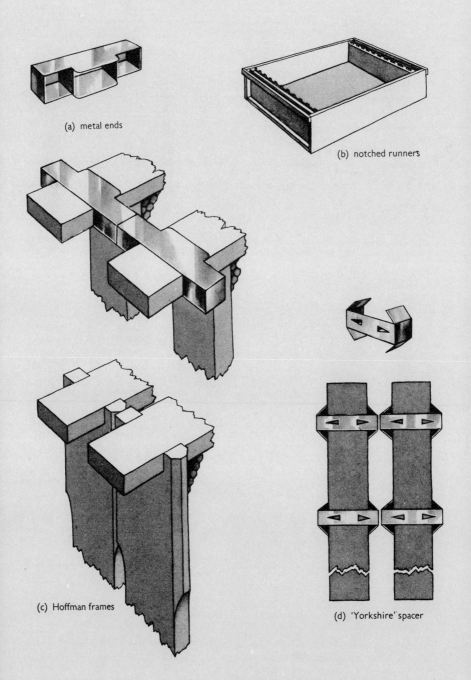

14. Methods of spacing frames

15. The W.B.C. hive

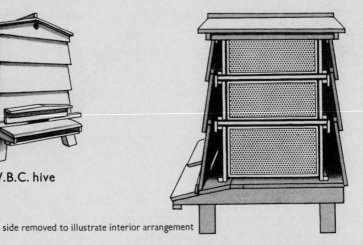

side removed to illustrate interior arrangement

16. The National hive

modified National hive

National hive
(roof now fitted with ventilators)

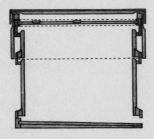

cross section of modified National hive

assemble these, nail the corners and fix the foundation. This procedure is easy but tedious; it is well worth while because ready assembled frames complete with foundation are expensive.

Spacing of Frames. Some means must be adopted of keeping these frames spaced at the natural distance of $1\frac{3}{8}$ inch–$1\frac{1}{2}$ inch from centre to centre or from one sheet of foundation to the next. The most usual (and the most inconvenient) method of doing this is to slip a special spacer $1\frac{1}{2}$ inch long and known as a *metal end* over each lug of each frame as shown in Figure 14 (*a*).

It has been found that if supers of drawn-out combs are spaced extra widely, say at 2 inch centres, the bees will deepen the existing cells and the honey capacity of a shallow frame is thereby increased from 3 lb. to 5 lb. Such a procedure leads to an economy in the use of frames and 7–8 can be made to do the work of 10. This method can only be applied to supers and in the brood nest the standard $1\frac{1}{2}$ inch spacing must be used. Moreover this method can only be applied to combs that are already drawn out. Bees will neglect foundation or build their combs between the frames if 2 inch spacing is used when frames are filled only with foundation. Metal ends are therefore made in two widths $1\frac{1}{2}$ inch and 2 inch, but the latter must be used with discretion. There is no reason why $1\frac{1}{2}$ inch spacing should not be used throughout the hive.

Metal ends are very widely used in spite of the fact that they trap bees between the lugs of the frames, interfere with the use of the knife when uncapping honey and are difficult to remove and replace because bees stick them firm with propolis.

Figure 14 (*b*) shows another simple method of spacing in which notches are cut in the metal strips which support the frame lugs. This is satisfactory in supers, but not in the brood chamber where the relative position of combs has to be altered from time to time.

Figure 14 (*c*) shows a type of frame in which the upper part of the side members of a frame are expanded to $1\frac{3}{8}$ inch or $1\frac{1}{2}$ inch. Such a frame is known as the *Hoffman type* and it is coming into greater use in spite of being more expensive. One edge of the expanded portion is generally sharpened or rounded so that the propolis seal may be more easily broken when frames are removed.

A very useful type of spacer known as the 'Yorkshire spacer' is shown in Figure 14 (*d*). This enables ordinary frames to be adapted to the Hoffman type of spacing.

HIVES

From the point of view of the bees a hive should be a weatherproof enclosure with ample room for brood rearing and food storage; a restricted entrance is also desirable so that the internal temperature may be controlled more easily and the home readily defended against enemies. From the beekeeper's point of view a hive should be easy to handle, extensible by the addition of extra boxes, standardized so that all parts of all similar hives are interchangeable, substantial, simple and cheap.

The design of the hives to take British standard frames has crystallized into three patterns—the 'W.B.C.' (after the initials of the inventor, W. B. Carr), the 'National' and the 'Smith' hives.

The W.B.C. Hive. This is shown in Figure 15. It is the oldest of the three types named and is still very popular. It is double walled, i.e. it consists of one set of boxes which serve to support the frames and to confine the bees; these boxes are made of thin wood and are quite inadequate to give protection against cold and moisture. Weatherproofness is obtained by the use of a second set of boxes constructed weatherboard fashion so that they entirely surround the inner set of boxes; the outer set of boxes are known as *lifts* and are constructed of $\frac{5}{8}$ inch or $\frac{3}{4}$ inch timber. This stout construction and the fact that there is an airspace between outer and inner boxes makes a very dry, warm and satisfactory hive. The inner boxes hold ten frames at $1\frac{1}{2}$ inch spacing. A gabled roof and a floor on four legs complete the hive.

The W.B.C. hive in addition to being snug is good looking and enhances the appearance of the garden. Its disadvantages are that due to its complicated construction it is expensive and owing to its double set of boxes it is rather cumbrous in use.

If expense is no object, if appearance is a consideration and if you are never likely to keep more than three or four hives you can scarcely do better than choose the W.B.C.

The National Hive. If your beekeeping is to be based on utilitarian and economical lines and if your activities are likely to extend to 10 or more hives you would be well advised to consider the National hive which is cheaper, easier to manipulate and yet quite satisfactory as a home for bees. This hive is of the single walled type. The boxes which support the frames are made of substantial timber so that they also give adequate protection against the weather. The boxes are square, holding 11 frames at $1\frac{1}{2}$ inch spacing and stand one on top of the other as shown in Figure 16. The cracks between the boxes are sealed up by the bees with propolis and this makes the hive weatherproof without the use of plinths or overlapping parts. A simple floor board and a flat topped roof covered with tarred felt or zinc sheeting complete the hive.

Paint on National hives tends to blister owing to the moisture generated by the bees themselves diffusing outwards and it is usual to treat such hives with creosote or wood preservative such as 'Solignum' or 'Cuprinol'.

The Smith Hive. This is the simplest and cheapest hive available yet quite as efficient as the two types previously described but its use demands a slight modification to the British standard frame. A glance back at Figure 12 will remind you that the lugs at each end of the top bar project $1\frac{1}{2}$ inch beyond the frame itself. These long lugs make the frames very convenient to hold but they do occupy a good deal of space in the hive and they require complicated shelves or ledges for their accommodation. If the length of this lug be shortened to $\frac{3}{4}$ inch sufficient support for the frame can be obtained and the appropriate bee-space maintained by cutting a ledge out of the thickness of the wooden end pieces of each box. A hive body can then be simply made from four pieces of wood joined at the corners.

Figure 17 shows the Smith hive in section and in view. It is made from $\frac{7}{8}$ inch timber and a ledge $\frac{1}{2}$ inch × $\frac{7}{8}$ inch is cut in the opposite ends. The brood chamber holds 11 frames at $1\frac{1}{2}$ inch spacing. The roof and floor are similar to those used in the National hive.

The Smith hive is the newest type of hive to appear on the market and is not so universally used as the other two types; nevertheless beginners would be well advised to adopt the Smith hive.

At the time of writing (1975) the relative cost of the three types of hive mentioned are W.B.C. £20·80, National £13·85, and Smith £12·40. A further saving in the cost of hives of 10% may be made by buying them in the flat and assembling them yourself.

Brood boxes for the $8\frac{1}{2}$ inch deep frames and supers for the $5\frac{1}{2}$ inch deep frames are available for all the hives mentioned.

HIVE ACCESSORIES

We have so far discussed types of hive and you may already have decided upon the type you want but there are one or two more points to consider before we get to the bees themselves.

Floors and Stands. Whatever type of hive we use it must stand on a level site. Nothing is simpler and better than a concrete slab 30 inches × 24 inches or 36 inches × 24 inches and this should be set level on the ground with the aid of a spirit level before the hive is placed upon it. Not only is this slab a solid foundation (bees hate vibration) but it prevents weeds growing close to the entrance of the hive. (See Figure 18.)

The floor of the W.B.C. hive is always provided with legs which raise it above the dampness of the earth. A stand for National or Smith hives may consist simply of four bricks or may be made of wood as the beekeeper may desire. Where the ground is very uneven, posts may be driven in and rails fixed as shown in Figure 18.

It is often necessary to restrict the entrance to the hive and in the W.B.C. hive this is accomplished by a pair of sliding strips of wood; in the National and Smith hives by a block of wood with slots cut in it. Both these devices are shown in Figure 19.

Inner Covers. The roof of the hive serves to keep out the weather and an inner cover must be provided over the top of the brood chamber or the top of the super as the case may be to retain the bees. A piece of stout fabric, e.g. deck-chair canvas, is a simple and popular method and is referred to as a *quilt*. Quilts have the advantage that you can turn them partly back and take out a frame or two without disturbing the majority of the bees. Their disadvantage is that they are likely to be bitten into holes, that they are not heat-retaining thus requiring extra packing on top especially in the winter and that, since bees stick them to the top bars of the top set of frames, there is no passageway over the top from one frame to another; in winter this is a serious drawback.

An alternative to the quilt is a *crown board*. This is a piece of wood the same size as the top of the hive; on its underside are nailed four strips of wood $\frac{3}{16}$ inch thick so that a permanent bee-space is provided over the top set of frames. A crown board may be made from moisture-proof plywood or from pieces of $\frac{3}{8}$ inch thick wood. Some makes of crown board have large glass panels so that the bees may be inspected without opening the hive; beginners will find these interesting and useful. (See Figure 20.)

All the above types of crown board require extra packing but if the board be made of timber of the same thickness as the hive—say $\frac{3}{4}$ inch—then no additional packing is required either in summer or winter. Unfortunately this type of crown board is not commercially available but it is well worth the trouble of making it for yourself. The disadvantage of crown boards is that they have to be removed completely from the hive so exposing all the top set of frames.

Both quilts and crown boards require a central hole about 3 inches in diameter known as the *feed hole*; when not covered by a feeder a piece of glass or perforated zinc is used to prevent the bees escaping.

Queen Excluder. This is a metal grid with slots of such a size that the queen cannot pass through while it offers little or no obstruction to the smaller worker. Figure 21 shows the two most usual types; (*a*) is a thin sheet of zinc perforated with long slots approximately $\frac{5}{32}$ inch wide. It is laid directly on the top of the bars of the brood chamber with its slots at right angles to the top bars of the combs; (*b*) is a grill of wires, spaced at the same intervals and mounted in a wooden frame. This provides a bee-space both below and above the excluder and it is not so necessary to place the slots at right angles to the combs. Both types of excluder serve effectively their purpose of keeping the queen out of the supers and thus confining her brood rearing activities to the brood chamber.

Smoker and Hive Tools. These two items of equipment are for the personal use of the beekeeper himself. One of the major discoveries in beekeeping is the fact that bees when subjected to smoke become more docile and are more easily handled; an efficient smoker is therefore of the utmost importance.

Basically a smoker consists of bellows upon which is mounted a cylindrical chamber to contain the smoke-producing fuel and this ends in a nozzle which serves to direct the resulting smoke. Air from the bellows blows through a small hole in the side of the fuel chamber so that, when the smoker is stood upright, the smouldering fuel can draw in enough air to keep itself alight but when the bellows are operated dense smoke comes (or should come) through the nozzle. Figure 22 shows two types of smoker that are in common use; (*a*) is the traditional British pattern but the fuel chamber is too small and has to be replenished every half-hour or so; (*b*) is the American pattern with a large fuel capacity so that it requires infrequent replenishing. It is also very convenient to handle. Smokers are made both in tinplate and in copper. The latter will last a lifetime and is well worth the extra money.

The handiest fuel for a smoker is a roll of corrugated cardboard but it burns away quickly and gives an acrid smoke. Old sacking, rotten wood or dried lawn cuttings all give a smoke which is more acceptable to the bees.

A cloth soaked in 'Izal' or other carbolic disinfectant and then dried is often used to subdue bees instead of smoke. It is placed over the top of the hive after the quilt or crownboard has been removed.

Since bees stick frames and hive bodies together with propolis a tool is necessary to prise them apart. A 6 inch screwdriver will serve the purpose but the special tool shown in Figure 22 is better and has one end designed for scraping away surplus propolis and brace comb. Choose a stout tool that will not bend under stress and have stainless steel if you can afford it.

We now have enough equipment to start beekeeping.

Suggestions for further reading

Butler, C. G. (1949) *Bee Hives*. Ministry of Agriculture, Bulletin No. 144. O.P.

Heath, L. A. (1986) *A Case of Hives*. Bee Books New and Old.

Wedmore, E.B. (1948) *A Manual of Beekeeping*, Sect. VII. Bee Books New and Old.

The following manufacturers of beekeeping equipment will gladly supply copies of their catalogues which beginners will find very informative.

E. H. Thorne (Beehives) Ltd, Wragby, Lincoln LN3 5LA.

Steele & Brodie, Wormit, Newport-on-Tay, Fife, DD6 8PG *and* Stevens Drove, Houghton, Stockbridge, Hants, SO2 6LP.

CHAPTER SIX

STARTING WITH BEES

ACQUIRING KNOWLEDGE

By buying, borrowing or begging this book you have taken an important step towards becoming a successful beekeeper. Books cannot teach you everything about beekeeping but it is cheaper and much more humane to gain preliminary knowledge by reading than by making mistakes and killing thousands of bees.

The next step is to make contact with beekeepers themselves so join the local Branch of your Beekeepers' Association; even though you have, as yet, no bees you will be welcome. The Agricultural Executive Committee of your County will willingly tell you the name and address of the secretary of the nearest Beekeepers' Association, or you may care to write to the IBRA, 18 North Road, Cardiff CF1 3DY, publishers of *Bee World*, who will forward your enquiry.

Beekeepers' Associations generally hold a series of lectures during the winter months and often run a special course of lectures for beginners. In the summer, outdoor meetings are held at members' apiaries and demonstrations using actual colonies of bees are given by experts. Not only are these summer meetings of the utmost value to beginners but they are a pleasant social occasion and give you an opportunity of meeting your local beekeepers and of inspecting well-run apiaries.

Old hands at beekeeping must on no account read the following.

While the established beekeepers are not listening let me warn you that the art of beekeeping has never been laid down on hard-and-fast lines and that there are, in consequence, almost as many systems of keeping bees as there are beekeepers, each of whom will argue plausibly in favour of his own particular method and the beginner will receive much conflicting advice. After the experience of a season or two you will have formed your own opinions on the merits and shortcomings of the various systems of management and you will be in a position to accept or reject the varying advice that is offered. But to the beginner it is very confusing indeed. Beekeepers are a friendly race however much they argue among themselves. They will be delighted to show you their apiaries and to demonstrate how bees should be handled.

Beekeeping Associations offer other facilities. They usually issue a monthly magazine free to all members; they have a team of experts who will give you a free visit and extricate you from the difficulties in which you are involved; they organize honey shows at which members vie with one another to exhibit the finest specimen of honey; they usually provide free third-party insurance against any injury that your bees may do to livestock or human beings, and for a modest premium they insure you against the Foul Brood Diseases which by law require the destruction of your stocks. For a modest subscription of about £2·00 per year this is wonderful value for money and you should join your nearest association without delay.

ACQUIRING APPARATUS

You will require the following apparatus:

    1 hive, consisting of one brood chamber and two supers.
    Roof and floorboard for hive.
    10 or 11 standard frames complete with foundation.
    20 or 22 shallow frames complete with foundation.
    1 spare hive consisting of brood chamber, roof and floorboard.
    1 Queen excluder.
    2 crownboards or quilts.
    1 smoker.
    1 veil.
    1 hive tool.
    1 pair gloves.
    1 feeder.

All the above have been dealt with in the earlier chapters excepting the feeder which is described on page 69.

A spare hive is always necessary for the accommodation of swarms though the need will probably not arise during your first year of beekeeping. Even if you have only one colony of bees you must still have a spare hive. If you keep more than one colony of bees then you will want a spare hive for every three or four stocks.

Whatever type of hive you choose, have the same type throughout your apiary so that every one is interchangeable; a mixed collection of hives is a sure source of trouble and unnecessary labour.

STARTING WITH BEES

One advantage of joining a Beekeepers' Association before you actually own any bees is the fact that the Secretary of the local Association is usually in touch with people who have surplus equipment to sell. Such equipment can often be obtained at about half the price of new equipment but if you decide to buy second-hand you must take the following precautions:

(1) Take an experienced beekeeper with you to make sure that the hives are of standard type. There are many obsolete types of hive still in use and beekeepers with a taste for woodwork produce hives of all shapes and sizes and to no standard at all.

(2) Make sure that the hives are sound and in a good state of preservation.

(3) Carefully sterilize all parts of the hive to which the bees have access by scorching them with a painter's blowlamp until the wood is light brown.

(4) Never purchase second-hand combs. They are particularly prone to harbour disease and it is not possible to sterilize them.

Get your apparatus and make yourself familiar with it before you buy your bees.

SETTING UP THE APIARY

If you are keeping bees in your garden set up your hives with their backs to the garden path. If bees come out of their hives and fly at low level across frequented paths frequent stings will be the result. If bees are flying at six feet or higher they are unlikely to collide with and sting anybody. They may be forced to rise into the air rapidly by erecting a screen in front of and three or four yards away from the hives. A hedge serves the purpose well. Jerusalem artichokes are effective in summer and $\frac{3}{4}$ inch wire netting, though not beautiful, is very efficient. Small gardens and near neighbours may force you to adopt some of these methods.

Since we shall not mention neighbours again it might be said here that a gift of a pot or two of honey seems to diminish the frequency with which your neighbours are stung; at least the number of complaints is noticeably reduced.

It is a good idea to place one's hives backing to the prevailing wind (in most districts this means they are facing east). Heavily laden bees can land more easily when flying into the wind.

Some beekeepers overcome the troubles of keeping bees in a small garden by installing them in the corner of some convenient field or orchard at a distance from home and such an arrangement is known as an *out-apiary*. An additional advantage lies in the fact that bees may then be situated in a better foraging district, among clover fields or fruit trees. The technique of managing an out-apiary is more difficult since the bees can only be visited at intervals so, unless circumstances compel you to adopt this method, wait for a year or two until you have gained experience.

ACQUIRING BEES

We have come to the great moment when we are about to become the actual possessors of live bees and must consider in detail how this is to be done. There are three ways of acquiring bees.

In the first place you may buy a stock of bees, i.e. a colony of bees complete in the hive which it inhabits.

Second, you may purchase a colony of bees established on combs and complete with a young laying queen and introduce it into an empty, hive. If the colony is small and the bees occupy only three or four frames it is referred to as a nucleus while in a colony the bees occupy 6, 8 or 10 combs.

Lastly you may buy, beg or catch a swarm and establish it in a hive fitted with frames and foundations.

Purchase of Stocks. Stocks of bees generally come on the market when the owner has died or is giving up beekeeping and the price is usually much less than the total cost of a hive and a colony of bees. On the other hand you will acquire a second-hand hive which may not be of standard pattern, the colony may be weak or diseased and the queen may be old. This method of acquiring bees is fraught with dangers for the beginner and should never be undertaken unless you can find an experienced beekeeper who is willing to give the stock a thorough examination. On his unbiassed recommendation you may purchase.

The problem of establishing the stock in your own apiary is simple and consists merely of closing up the hive and transporting it and its contents from its old situation to the site you have prepared for it. If the distance moved is over a mile there is little danger of the bees returning to the original site of their home.

Purchase of Colonies and Nuclei. Most appliance manufacturers also supply colonies of bees for sale. In addition many beekeepers specialize in the production of colonies and their advertisements may be found in the bee press. If you have already joined a Beekeepers'

Association one of the members may be a breeder and in this case you will be able to visit him, see the bees that you are buying and possibly take your purchase home with you in a car.

When deciding whether to buy a nucleus or a colony bear in mind that the price (1975) varies from £3/£4 per comb so that you may pay from £10 to £25 for your bees but you must set against the higher price the fact that the larger colony will probably produce a crop of honey in the first year that will compensate for the extra money spent. The prospect of having your own honey at the end of your first year of beekeeping is a strong inducement to purchase a colony.

A nucleus will spend all its energies and use up most of its stores in expanding and filling up the brood chamber of your hive. You are very unlikely to get surplus honey from it during the first year but you will end the season with a strong colony covering 10 or 11 frames and in good condition for facing the coming winter.

A colony (or nucleus) should fulfil the following conditions:

(1) It must be healthy. If your supplier hesitates to give you a written guarantee to that effect, take your order elsewhere.

(2) The colony should contain a laying queen not more than one year old.

(3) The bees should be docile and should occupy all the combs.

(4) Not less than two thirds of the comb area should be occupied by brood and eggs.

Colonies are available from early May until August but the earlier they are obtained the better. Your colony will arive in a special lightweight bee-proof hive known as a *travelling box* (see Figure 23). It has gauze or perforated zinc panels so as to provide ample ventilation for bees get very excited and hot when travelling and may suffocate. They are generally despatched by passenger train and at the same time a telegram will be sent by the supplier so that the bees may be collected without delay.

Set the box on the ground close beside the empty hive and then remove the strip of perforated zinc that covers the entrance. The loud buzzing inside the box shows that the bees are very excited and they will pour out of the entrance and fly wildly around the moment you release them. Since you are only just beginning beekeeping it is better to use veil and gloves even for this simple operation.

By the next day the excitement will be over and they may be transferred to the hive. Open your hive and have your brood chamber empty. Don veil and gloves and puff a little smoke into the top of the travelling box. Wait two minutes and then remove the lid. Very gently and deliberately transfer the combs and adhering bees, one by one, into the hive, keeping them in exactly the same order. When this has been accomplished fill up the remaining space in the brood chamber with frames containing foundation, equal numbers on either side of the colony.

The brood chamber may be placed so that the top bars of the frames are parallel to the entrance and this arrangement is known as the *warm way*. Alternatively the frames may be at right angles to the entrance and this is known as the *cold way*. It has been shown that there is no difference from the bees' point of view but the cold way seems to me to make for easier manipulation.

When you have replaced the quilt or crown board it is a wise precaution to give your colony a feed of sugar syrup (see page 68 for details). A few bees will remain clinging to the sides of the travelling box. Place a sheet in front of the hive and shake them out on to this; they will soon find their way in and join the rest.

Purchase of Swarms. A swarm consists of an old queen together with 15,000 to 30,000 workers. Swarms are usually sold by weight and you should try to get one weighing not less than 4 lb. (4,000 bees per lb.).

The price of swarms varies with the month as the old rhyme beginning

'A swarm of bees in May
Is worth a load of hay . . .'

indicates. A swarm of bees in May will probably cost about £2·00 per lb. but it will rapidly establish itself and, in a good season, will give a good surplus of honey. In July, if you know several beekeepers, you can often have a swarm as a free gift, but it will require to be fed liberally to establish it for the winter.

The advantage of a swarm is that it is the cheapest way of acquiring bees. The disadvantages are that there is no way of telling whether a swarm is healthy; the bees may come of a swarming strain and thus cause endless trouble and it is difficult to judge the disposition of bees under such circumstances. Nevertheless many hundreds of beekeepers have been started by the gift of, or the lucky discovery of a swarm.

Bee breeders do not sell swarms so you will have to obtain it locally. The bees will be in a box or *skep* covered with muslin or in a travelling box. They will be without combs although, if they have been in their container for more than 24 hours, they will have started to build natural comb.

The sooner the swarm is hived the better. Open your hive and fill the brood chamber with frames fitted with foundation. Replace quilt or crown board and close up the hive again. Spread a sheet in front of the hive and see that one edge of it comes close to the entrance so that the bees can march in without having to cross any gaps. Take the box containing the bees and holding it with the opening downwards shake out the bees with a sharp downward jerk on to the sheet close to the entrance to the hive. In less than a minute they will start to march in and within half-an-hour the occupation of the hive will be complete. After hiving, a swarm must be fed liberally for several days. Figure 24 shows a swarm being hived.

At last you are actually keeping bees.

Suggestions for further reading

Magazines regularly carrying advertisements for stocks, colonies and equipments, include:

The British Bee Journal. 46 Queen Street, Geddington, nr Kettering, Northamptonshire.

Bee Craft. 17, West Way, Copthorne, Sussex.

The Scottish Beekeeper. Schoolhouse, Kirkpatrick Fleming, Lockerbie, DG11 3AU, Scotland.

CHAPTER SEVEN

PLAN OF CAMPAIGN

Now that you are the actual owner of one or more stocks of bees you must consider what attention they require during the course of the year.

Almost certainly you will give them too much attention to begin with; you will open the hive too frequently just to see how they are going on or to make sure that the queen is alive and well. In your first year this is not altogether bad because you must get to know your bees, their way of life and the details of the home that they make for themselves. Apart from curiosity it should not be necessary to open up and go through a hive more than twice in a season unless that colony swarms. To begin with you will probably open your hive five or six times during the season but always remember that bees that are continually disturbed tend to become bad tempered and that some of your precious honey crop is lost each time you open up on account of the disorganization caused.

THE BEEKEEPER'S YEAR

Both bees and beekeepers become active in the spring and this makes a good starting point for a survey of the year's work. To begin with we shall consider a colony that goes through the season without swarming, leaving swarming with all its associated problems until the following chapter.

In beekeeping, even more than in gardening, it is difficult to quote dates at which operations should be carried out. Spring is the most critical season for bees and it is also the most variable and unpredictable from the point of view of weather. Much more reliable than the calendar as a guide to beekeeping operations is the time of appearance of certain flowers. These times do not have the hard inflexibility of calendar dates; they take into account the conditions of the previous months very much as do the bees themselves.

Here then is our programme. Certain operations which would occupy too much space to deal with in our calendar are described at the end of this chapter.

Crocus in Flower. Note whether there is activity on sunny days. If your bees are carrying in pollen this is the best of all signs but do not be discouraged if there are no signs of activity; some stocks are noticeably late in starting but these often turn out to be unusually good. You may put your hand directly on the crown board or quilt and, if you feel signs of warmth, you will know that spring breeding has begun and that the queen is alive.

Put your hand under the back of the hive and lift it about two inches so that it rests only on its two front legs or the front edge of the floor board. (Figure 25.) A little practice will enable you to judge by weight the quantity of stores present. If they appear short of stores feed with candy or syrup.

Plum Trees in Flower. On a warm day carefully lift the complete brood chamber from the floor board and set it aside on the upturned lid or on one of the lifts if the hive is of the W.B.C. type. Remove floor board and replace it by a clean one. Replace brood chamber and rebuild hive but on no account open or disturb the brood chamber.

Check stores again by lifting hive and feed if necessary.

Apple Trees in Blossom. This is the time for a first thorough examination of the brood chamber. Wait for a day that is warm enough for you to take off your jacket out of doors without feeling cold. Smoke the hive gently (see page 66) and then remove the cover.

The objects of the spring examination are threefold: (1) to confirm the presence and activity of the queen, (2) to determine the state of the brood chamber, and (3) to detect disease.

(1) It is not necessary to see the queen herself to confirm her presence and her activity. We judge her by the results of her labours, by her eggs and brood. If her eggs are found the queen was certainly present three days ago and it is fair to assume she is still alive unless your careless manipulations have crushed her. So look for eggs first of all. Then note the brood both sealed and unsealed; there should be brood on about six frames by this time of the year, the biggest area of brood being on the central frame. The brood should be in continuous slabs with very few empty cells; a queen who lays in an erratic irregular fashion is old or unsatisfactory. At this time of the year there should be very little drone brood; excessive drone brood is a sign of an old or failing queen.

(2) Note first the number of frames containing brood; if all but the two end frames contain brood, the chamber will be overcrowded within a week or so, and in this case give the queen more space by adding

to the brood chamber an extra box of frames. Next note the amount of stores remembering that a brood frame if full, will hold about 5½ lb. of honey. For the first time you have an opportunity accurately to check the amount of stores present; if there is less than 5 lb. or if the weather is continuously bad, feed because this is the most critical time for the colony. Note whether there are cells containing new nectar. There may even be capped spring honey cells detectable by the fresh whiteness of the capping.

Also take note of the condition of the frames themselves; if any of them are nibbled away or show holes to the extent of 20% of their surface area, they should be replaced at once if unoccupied or, if occupied, should be placed on the outside position in the brood chamber so that they may be removed on some future occasion when the brood nest has contracted and left them unoccupied.

(3) Last, look at the brood itself for signs of disease. These will be discussed more fully in Chapter 11 so let us content ourselves by saying that the signs to keep an eye open for are: unsealed brood that displays any suspicion of brown colour; sealed brood that shows signs of sunken or pierced cappings (unless the fully developed bee is engaged in biting its way out); unsealed brood that looks either black or chalky white. Call in an expert if any of these signs are observed.

If nectar is being brought in freely it is a good idea to put on a queen excluder and super when rebuilding the hive.

Hawthorn in Flower. This is the beginning of the swarming season. Your main job is to see that the hive never becomes overcrowded with bees or overfull of nectar and honey since either of these conditions provokes swarming. Once every ten days open the hive and glance quickly into the super to make sure that there is ample storage space, i.e. empty cells for nectar. Add another super if 70% of the available space is occupied. If you have drawn-out combs it is a good idea to make these your first super. The second super may well be frames filled with foundation. The advice generally given is that the second super should be placed underneath the first, but I have never found any advantage from this procedure so I should save yourself trouble and simply place the second super on top of the first.

At the same interval of ten days examine the brood chamber. If this is still a single box containing 10 or 11 frames it will almost certainly be full by this time. When brood appears on the penultimate

17. The Smith hive

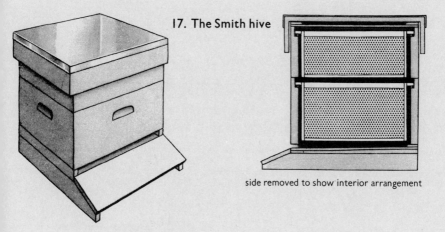

side removed to show interior arrangement

18. Stands for single-walled hives

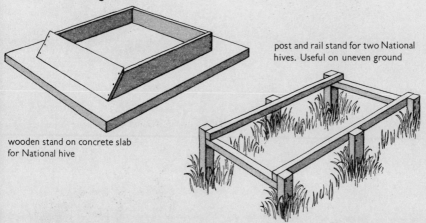

wooden stand on concrete slab for National hive

post and rail stand for two National hives. Useful on uneven ground

19. Slides for restricting entrance on W.B.C. hive

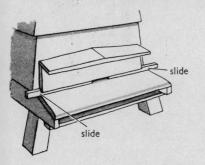

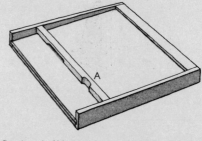

floorboard of National hive with entrance block in position. If block is turned over the smaller slot (marked A) comes into use and the entrance is still further restricted

20. Crown boards

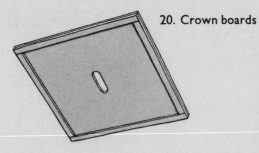

crown board (seen from below)

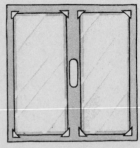

crown board with glass panels

21. Types of queen excluder

(a)

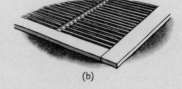
(b)

22. Types of smoker, and a hive tool

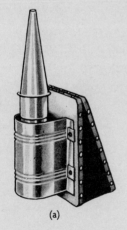
(a)

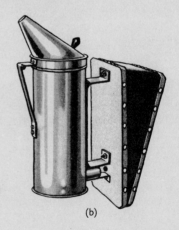
(b)

hive tool

frames add a new box of frames to give the queen extra room. These extra frames should preferably be already drawn out but in your first year of beekeeping this will not be possible and you will have to be content with frames of foundation. An extra box of shallow frames will give most queens ample room but you may add a box of deep frames on top of the original brood chamber if you wish. Such an arrangement by which the queen is given access to two boxes of combs is known as a *double brood chamber*. All this will be done with very little disturbance to the hive.

While examining the brood chamber for overcrowding it is a good thing to carry out a cursory examination for queen cells so that one may be forewarned of imminent swarms. (See Figure 26.) If your bees are already on a double brood chamber it is only necessary to examine the lower edges of the top set of frames for this is a favourite situation for queen cells. If your bees are on a single brood chamber, examine the three centre combs with the minimum of disturbance. If queen cells are found, your course of action is laid out in the following chapter.

Lime Trees in Flower. This is the peak of the season for the suburban beekeeper. The danger of swarming is now largely over but keep a careful eye on the capacity of the supers; the nectar flow from the limes is often very heavy and in these circumstances the bees may fill a super in a few days.

Willow Herb Finishes Flowering. This is the end of the nectar flow for all beekeepers except those with access to heather. In the midlands this usually occurs about the first week in August and this is the time to remove the supers.

The nectar flow ceases abruptly and as a consequence many foraging bees are thrown out of work; these may attack weak hives and rob them of their honey. As a precaution therefore restrict the entrances to your hives so as to give the rightful owners a better chance of defending their property. You will remember that Horatius with two companions was able to defend Rome against the whole Tuscan army under similar circumstances. We shall have more to say about robbing under 'Troubles' in Chapter 11.

Montbretia in Flower. Now is the time to prepare for the coming winter. Since the supers have been removed your hive is reduced to a brood chamber once more. If the colony is strong it may be allowed to retain its double brood chamber throughout the winter. If weak it is better to reduce it to a single brood chamber again, either by removing unused combs or by putting the box of shallow frames which has hitherto been the top of the original brood chamber underneath. The bees will migrate upwards, taking with them any stores that may be present and, at about the time of the first frosts, the lower box may be removed.

Early in September go carefully through the colony, lifting up each frame and assessing the amount of stores each contains. If the total amount of stores reaches or exceeds 35 lb. that colony will not need further feeding for the winter. If the total falls short of that figure then the quantity must be made up by feeding the necessary weight of sugar made up into thick syrup (see page 68). If you can afford to make up the total stores to 40 lb. so much the better. All your feeding operations should be over by the 1st of October.

Your hive entrances should be restricted during the feeding period for robbing is still very prevalent and wasps are troublesome. During the winter it is undesirable to restrict ventilation and hive entrances should be opened to at least 5 inches. On the other hand some restriction is still very desirable since field mice are fond of spending the winter inside the hive. The best plan is to fasten with drawing pins a piece of perforated zinc across the entrance and to cut an aperture 1 inch × ¼ inch in this to give the bees means of coming and going.

Check your hive to make sure it is weatherproof; the roofs of W.B.C. hives are very prone to leak and need special care. See that the roof of your hive is not likely to be blown off in a gale. W.B.C. roofs are vulnerable and are best secured with a cord.

During the winter period from October to April the hive must not be opened nor the bees disturbed.

That completes the cycle of the beekeeper's year (swarms excepted) but we must consider in more detail one or two of the operations referred to.

OPENING A HIVE

First gather together all the tools and equipment you are likely to need. Then light your smoker and, when it is fairly alight, put on your veil and get to work. Very little smoke is necessary. The object of smoke is not to stupefy or half-asphyxiate the bees but to inject into the hive a strange smell which sends them running to the nearest open cell; from this they will fill their honey sacs, so providing themselves with about three days' emergency rations should the disaster take place which the strange frightening smell of smoke seems to them to forebode; and a bee with a full honey sac is a good-tempered bee. You will also see that it is no use smoking a starving colony for smoke can only subdue bees when they have access to stores. Three or four puffs of smoke are sufficient. Many beekeepers blow the smoke into the entrance of the hive but it is much better to lift a corner of the quilt or crown board and blow smoke into the top of the hive. After smoking wait about two minutes before removing the cover so as to give the bees time to fill their honey sacs.

This is the moment to emphasize once again, the necessity of slow, deliberate movements when handling bees. Remember also that you should never stand in front of the hive in the line of the flying bees when examining the stock. If the cover is a fabric quilt, begin at one corner and strip it off, breaking the propolis seals with a steady, deliberate pull. If you have a crown-board, lever it up slightly with the hive tool, twist it through a few degrees to break propolis and brace-comb seals and then lift off slowly but steadily. Lay the quilt or crown-board upside down in front of the hive so that the bees still adhering to it are not crushed.

Since ten or eleven frames fill a brood chamber, the removal of the first frame is awkward. If your brood chamber is arranged the cold way, the two end frames will be the most thinly populated and least likely to contain the queen. With the hive tool, break the propolis seal and raise one of these slowly, taking care not to crush any bees.

Holding the frame by its lugs with the top bar horizontal and the comb hanging down one can only inspect one side. To look at the other side one must turn the frame over without bringing the comb into a horizontal position. Nature never designed combs for the horizontal position and if they are held in this fashion they are likely to bulge or break and the contents of the cells (especially nectar) may fall out. Figure 27 shows the three simple movements which enable you to turn a frame over without the comb becoming horizontal.

When you have looked at both sides of the first frame it is a good idea to shake the bees off and to lean the frame against the hive so that you have comfortable working space within the brood chamber for withdrawal and examination of the remaining frames. To knock bees off a frame let it hang in a vertical position, grasp the upper lug firmly and with a clenched fist sharply strike the hand holding the frame. The resulting jerk will detach most of the bees who seem too surprised to resent the action. Do this operation over the alighting board so that the bees may easily find their way home again. See Figure 28.

Examine the remaining frames of the brood chamber, loosening each frame with the hive tool before attempting to lift it out of the hive. It will not be necessary to knock the bees off the remaining frames unless a very minute examination of the contents of the cells of a particular frame is necessary. It will be found that bees cling to their comb after a hive has been smoked unless deliberately dislodged. After examination replace each frame in the brood chamber and when rebuilding the hive take care that the frames are put back in their original order

FEEDING

Nothing is more demoralizing to a colony of bees than the prospect of starvation. To guard against such a contingency it is necessary to feed bees on certain occasions.

The most important feeding time is the month of September when it is necessary to ensure that the bees have sufficient stores—at least 35 lb.—to last them through the winter.

Some beekeepers also feed their bees in the spring with the idea of simulating an early nectar flow and encouraging the bees to begin breeding earlier than they would normally. It is very doubtful if this practice has any advantages unless there is an unusually cold and prolonged spell of weather in the spring.

Whenever a new colony of bees is being established or a swarm is building a new home for itself, feeding will be of immense help and should be resorted to even at the height of the season.

In an emergency, whenever a stock is found to be dangerously short of food (and owing to our variable seasons this might happen at any time of the year but is commonest in April), they must be fed to save them from starvation.

Feeding is generally carried out by dissolving sugar in water to produce a syrup. The generally recommended strength is 2 lb. of sugar to one pint of water but bees seem to prefer it a little weaker than this and I use 5 lb. of sugar to 3 pints water. The sugar is dissolved in hot water but it is not necessary to boil the solution. If you use three pints of water to 5 lb. of sugar you will end up with 5 pints of syrup since the volume of water increases in dissolving the sugar; this is convenient since 1 pint of syrup of this strength contains 1 lb. of sugar and you may make up your syrup in bulk and judge the amount of sugar given by the volume fed. Beekeepers who practice spring feeding generally employ a thinner syrup made by dissolving 1 lb. of sugar in 1 pint of water.

Older textbooks stress the desirability of using cane sugar for feeding bees but wartime experience, when beekeepers used whatever sugar was available, did not indicate any difference between cane sugar and beet sugar.

Various additions to the sugar syrup have been suggested from time to time but only thymol gives any definite advantage. Thymol is an antiseptic and has the power of preserving sugar solution against mould and fermentation indefinitely; if bees are slow in taking their feed (as they are in cold weather) syrup containing thymol will not spoil. Thymol is not strange or distasteful to bees since it occurs naturally in nectar from plants of the mint family. The recommended quantity of thymol is two-fifths of a grain per pound of sugar used. A convenient way of measuring this is to get from a chemist a 2% solution of thymol in surgical spirit and to add one teaspoonful of this to every pint of syrup.

Bees regard sugar syrup as nectar; an obvious way of inducing them to take it would be to set a dish full of syrup in the open air not far from the hive. Occasionally this is done by lazy beekeepers but the plan has grave disadvantages; you have no control over the amount taken by each individual hive; you probably provoke an outburst of robbing; you feed your neighbours' bees as well as your own; you feed the local wasps and other insects in which you have no interest. For these reasons it is usual to use an individual feeder for each hive.

A very simple and effective feeder may be made by taking a lever-lid tin such as a Lyle's Golden Syrup tin and making about a dozen small ($\frac{1}{16}$ inch) holes in the lid. The tin is then filled with syrup and inverted over the feed hole. The holes in the tin are too small for the thick syrup to run through; on the other hand a bee can stick its tongue through the hole and suck out the syrup. The same principle may be applied to a 14 lb. honey tin which holds a gallon of syrup and with this extensive feeding can be carried out very quickly and easily. A feeder on a similar principle but with a device for varying the number of holes available to the bees is on the market but its syrup reservoir is a 1 lb. honey jar and its capacity is so limited that it is little more than a toy.

Figure 29 shows an effective type of feeder which is available in 1½ and 4 pint sizes. Bees come through the feedhole and climb up the centre tubular hole of the feeder which is lined with perforated zinc to provide foothold. Surrounding the tube and partly immersed in the syrup is a wood cone which gives the bees firm footing while they are taking in the syrup. A glass topped cover prevents the bees leaving the hive through the feed hole and a lid over all keeps out dirt and robbers. The great advantage of this feeder is that it can be refilled without releasing or disturbing the bees.

When feeding it may be necessary to provide extra headroom above the crown board so that the roof can be replaced over the feeder. An extra outer lift on the W.B.C. hive or an empty super or brood box on a National or Smith hive will do the trick.

PUTTING ON AND REMOVING SUPERS

It is not usually worth lighting a smoker for these operations and a carbolic cloth is a great help. When using the cloth, remove the quilt or crown-board gently and without disturbance and immediately cover the top of the hive with the cloth. Or it is possible to lay the cloth on to the top of the loosened crown-board and then to withdraw the latter from under the cloth so avoiding the exposure of the unsubdued bees even for a second. Wait for two minutes and then withdraw the carbolic cloth. The bees will have all been driven downwards and the exposed top of the hive will be practically free of bees. The bees will remain below for at least half a minute which gives plenty of time to put on the empty super. It is better to put down one edge of the super first and then to lower it slowly into position so as to give any stray bees who may be standing on the edge of the brood chamber a chance to escape without being crushed.

Of course smoke may be equally well used if you consider it worth while to light the smoke for an operation that can be carried out in one or two minutes.

The removal of the supers is made very easy by a piece of apparatus known as a *clearer board* and this is shown in Figure 30. It is a piece of thin wood the same size as the top of the hive with a wooden fillet all round the edges of both sides. In the centre of this board is a hole into which is fitted a device known as a *Porter bee escape*. Bees passing through the escape have to push past a pair of weak springs which close behind them and prevent their return. A clearer board usually contains a second hole covered by a metal slide that can be operated by a projecting lever. If this hole is uncovered the bees are readmitted to the supers.

Removal of supers should be carried out in the evening when bees have ceased flying and is made much easier if an assistant can be obtained. Lift off all the supers together and stand them on a lift or the upturned lid of the hive; this is where assistance is required if the season has been a good one. With a carbolic cloth or a little smoke drive the bees from the top of the brood chamber, remove the queen excluder and replace it by the clearer board. Great care must be taken to see that the board is the right way up and it is advisable to paint the word 'top' on that side of the board which should be uppermost. The supers are then replaced on top of the board.

In 24 hours, if all is working well, the supers should be clear of bees. If not, investigate the following possibilities:

(1) The springs in the bee escape are bent too far apart so that bees can pass in both directions.

(2) The escape is blocked up by dead bees.

(3) There is brood in the supers because the queen has managed to squeeze her way through the queen excluder earlier in the season. Bees will never desert brood and any frames containing brood must be taken out and kept in a bee-proof place until ready for extraction, having first knocked or brushed off the bees.

(4) There is some external hole or crack in the hive by which the bees may re-enter the supers.

When the supers are free of bees they may be taken off and stored in a bee-proof place until the process of honey extraction is undertaken. At this time of the year the foraging bees are on the prowl and, if they can get access to the supers which you have removed they will quickly carry back the honey to the hive and your labour will have been in vain. In order to avoid raising these predatory instincts by exposing honey, all operations connected with the removal of supers should be carried out in the evening when no bees are flying.

Suggestions for further reading

Manley, R. O. B. (1948) *Bee-keeping in Britain*. Faber and Faber. O.P.

Digges, J. G. (revised Manley, R. O. B.) (1948) *The Practical Bee Guide*. Simpkin Marshall Ltd. O.P.
Phillips, E. F. (1943) *Beekeeping*. Macmillan. O.P.
Butler, C. G. (1946) *Beekeeping*. Ministry of Agriculture Bulletin No. 9 (11th edtn 1972 by F. W. Shepherd).
Wadey, H. J. (1944) *The Bee Craftsman*. 'Bee Craft' Books. O.P.

CHAPTER EIGHT

SWARMS

If the methods described in the previous chapter seem complicated it is only because I have described in detail the fundamental operations of beekeeping. You should find them easy to carry out, particularly if you have had the opportunity of watching an experienced beekeeper beforehand. The whole art of beekeeping would be simple if only bees did not swarm.

Many attempts have been made by selective breeding to produce a non-swarming strain of bee and some of them have been fairly successful. Unfortunately, such bees show deficiencies in other directions and are lazy or unprolific or poor nectar gatherers so that the best we can do in this direction is to try to acquire a strain that is not excessively prone to swarming since bees show a marked variation in their tendency to swarm.

AVOIDANCE OF SWARMS

Certain conditions within the hive tend to promote swarming and we must do all we can to avoid these conditions.

Overcrowding. If ever the queen has to search around for an empty cell in which to lay her next egg the brood nest is inadequate and overcrowded. Once an egg has been laid in a cell, that cell will be occupied for at least 21 days; if a queen is laying at the rate of 1,500 eggs per day she will require $1,500 \times 21$ or 31,500 cells during the peak of the laying season. Ten standard brood frames contain about 50,000 cells, but a queen will not use the end frames unless overcrowding compels her to do so and the bees will use some of the brood cells for the storage of honey and pollen. So do not hesitate to provide the queen with a double brood chamber long before she begins to feel short of laying space. Later in the season it is equally important to see that there is adequate space in the supers for all the nectar coming in, otherwise the bees will store excess nectar in the brood chamber and the queen may not then find room to lay even in a double brood chamber. No harm is done in putting on supers too soon, but the whole organization of a colony may be upset by putting them on too late.

Age of the Queen. A colony headed by a queen in her first year will rarely swarm but as she grows older the chances of a swarm increase. Some beekeepers suppress swarming by requeening each colony every year but this is not only laborious but uneconomical since queens are usually at their best during the second year. Replace any queens that are over two years old or are unprolific.

Presence of drones. Keep drone production to a minimum by replacing any combs in the brood chamber which show an excess of drone cells. The brood chamber as a whole should contain less than 5% of drone cells. Since drones are necessary for mating a new queen, excessive numbers seem to inspire the bees to raise queen cells and queen cells nearly always lead to swarms.

Weather. You have no control over this but it seems worth mentioning here that swarming is much more prevalent in some years, generally those in which the nectar flow is intermittent owing to variable weather.

TAKING SWARMS

Even though you take all the above precautions a time will come when, about midday on a warm sunny day, a swarm will emerge from your hive and you will have to cope with it. Turn back to p. 36 and read once again the events which take place when a swarm leaves the hive.

If you actually saw the swarm come out then you must wait until it has clustered in its first temporary clustering place; this process may be speeded up by spraying the flying bees with water from a garden syringe. If the cluster is in an accessible place it is only necessary to shake it into a box, cover the box with a cloth to retain the bees and then to place the inverted box on the ground in the shade not too far from the clustering spot. Prop up one edge of the box with a stone so that the bees can come and go. Inevitably a small part of the cluster will take wing and not be caught in the box but if the queen is among the bees in the box, the rest will join her within an hour or two.

Often a swarm will cluster in an inaccessible place, (e.g. within a thick hedge), or in a place from which they cannot be shaken (e.g. a wall or the trunk of a tree). In such a case the best plan is to fix a box mouth downwards above the cluster. The bees' tendency is to move upwards and in doing so they will probably occupy the box. This tendency may be accelerated by the discreet use of smoke or carbolic cloth at the lower end of the swarm. The box will attract the bees more strongly if it is baited by fixing in it a used comb or better still a comb containing a little brood. On the following morning the bees will be found to have moved up into the box.

HIVING SWARMS

Method 1. An old and obvious method is to return the swarm to the hive from which it issued. This is a bad plan because it restores the state of affairs in the hive which originally caused the swarm to issue and the swarm will certainly reissue, probably on the following day. This reissue may be delayed if the queen cells are removed from the parent stock but it will probably take place in the end.

Method 2. Another very obvious method is to hive the swarm in a new hive on a new site. This works well with a strong swarm during a nectar flow but, if the weather turns bad, the newly established swarm must be very heavily fed since it has no stores of its own.

A swarm is filled with desire to make wax and, when preparing a brood chamber to receive a swarm, frames of foundation are used so that the bees may satisfy their instincts by cell building. Nevertheless it is a good idea to put one drawn-out comb in the centre of the brood chamber so that the queen may begin laying without waiting for the workers to complete the first cells.

The most usual way of introducing a swarm to its new home is to spread a sheet in front of and right up to the entrance to the hive. The box containing the swarm is then carried to the prepared hive and with a sharp jerk all the bees in the box are thrown out on to the sheet as close to the entrance as possible. A few bees enter and start emitting their characteristic scent; in a very short time all the bees have turned their heads towards the hive and are marching in and within half an hour nothing will be left outside except a few dead or injured bees. (See Figure 24).

This method is useful where stray swarms have to be hived but if it is used on a swarm that has emerged from the beekeeper's own apiary then the brood chamber of the parent stock must be examined and every queen cell except one must be destroyed so as to prevent the colony from throwing casts. It is advisable to inspect the parent colony a second time after an interval of six days to make sure that the bees have not raised a second crop of queen cells from eggs or young larvae existing in the hive.

Method 3. By far the best method of dealing jointly with the swarm and the parent stock is that devised by the old British beekeeper Pagden and named after him. Pagden's method takes place in four stages shown in Figure 31.

First, discover which hive (if you have more than one stock) contains the parent colony. Remove the complete hive to a new site at least two yards away from the old site.

Second, place a floor board and a new brood chamber prepared as described under method 2 on the old site.

Third, take the supers together with the bees they contain and the queen excluder and place them on the new brood chamber which stands on the old site.

Fourth, run the swarm into the new brood chamber.

Bees memorize the site of their hives and not the hive itself; they will enter any hive on the site to which they are accustomed to return. When the Pagden method has been carried out therefore all bees flying from the parent stock which was set aside will return to the old site and augment the swarm. The augmented swarm will continue to accumulate honey as fast as did the original colony, faster perhaps, since temporarily they have no brood to feed and since the act of swarming seems to have a vitalizing effect on a colony.

The parent colony may be dealt with in several ways. It cannot swarm since it is denuded of flying bees and it will itself dispose of its surplus queen cells and raise only one queen; if you do not trust it to do this you may go through the colony and destroy all the queen cells except one. Ultimately the new queen will mate and the parent colony will become a new stock headed by a new queen. The act of swarming has increased the number of your colonies by one. If a new queen is desired without increase of colonies then the parent colony may be united (see Chapter 9) to the swarm at the end of the season. It is better to remove the old unwanted queen before doing this.

If more than one new queen is wanted the parent colony may be broken up into two or three nuclei each containing one queen cell. These nuclei are not self-supporting and have to be fed but each will produce a fertile queen if well looked after.

If no increase and no new queens are desired then all the queen cells should be destroyed and in six days a second examination should be made and any further queen cells destroyed. After this the queenless parent stock may be united to the swarm without fear of provoking further swarms.

DETECTING AND DEALING WITH IMPENDING SWARMS

If when examining a brood chamber the beekeeper notices a number of queen cells this should be a warning that a swarm is impending. When a hive is preparing to swarm it will build between 5 and 20 queen cells around the edges of the frames and these will vary considerably in the age of their occupants. It should be noted however, that bees also build queen cells when preparing to *supersede* their queen but in this case there will be only one or two cells on the face of the comb and they will generally appear before or after the normal swarming season.

If all the queen cells are still unsealed you may be certain that the swarm has not emerged for it will not leave the hive until at least one cell is sealed. If at least one queen cell is sealed but the old queen is still present, the emergence of the swarm is imminent. If at least one queen cell is sealed and the old queen is not to be found then the swarm has emerged.

Since it takes about nine days from the laying of the egg to the sealing of the queen cell, a beekeeper who gives a quick glance into the brood chamber once every ten days as suggested on page 64 will be warned of swarms before they emerge and will be able to take action to prevent them.

Method 1. An obvious method is to break out and destroy all the queen cells. Occasionally, if the cells are discovered at a very early stage, their destruction will discourage the bees from making any more. In general, though, the bees set about producing a fresh set of cells and if these are destroyed the bees become bad-tempered owing to the continued baffling of their instincts. Moreover they tend to build their cells in concealed places and, if even one cell is overlooked and escapes destruction, the whole method falls to the ground. This method is bad and should only be used in exceptional circumstances.

Method 2. Set the brood chamber containing queen cells upon a new site at least three yards from the original site. Upon the original site set a floor board and an empty brood box and upon it place the supers. Bees of flying age coming from the brood chamber will return to the old site and enter the supers and this goes on until the brood chamber is stripped of its flying bees and swarming is impossible. When the bees realize this they will of their own accord tear down the queen cells and abandon the idea of swarming. In seven days the brood chamber can be replaced on the old site under its supers, having first removed the empty box. It is not necessary to inspect the brood chamber. This method is very simple and effective but the bees in the supers, divorced from their brood chamber and queen, tend to become bad-tempered, although this is only temporary.

Method 3. While it is rather more laborious and requires a spare hive, the operation known as *Artificial Swarming* is the best way of dealing with a hive that is about to swarm. It bears a marked resemblance to the Pagden method described earlier in this chapter.

First remove the whole hive to a new site at least two yards away from the old site.

Second, place a new brood box on the old site. This should have 10 frames of foundation if of the National or Smith types or nine frames if of the W.B.C. type, thus leaving space for one extra frame. Arrange the frames so that this space lies in the centre of the brood chamber.

Third, remove supers from the original hive and, using a minimum of smoke, go carefully through the brood chamber until the frame containing the queen is discovered. Take this frame, together with the adhering bees, and place it very carefully in the vacant space left in the new brood box.

Fourth, place queen excluder and supers on top of the new brood chamber and rebuild the hive. Add a comb—either foundation or drawn-out—to the old brood chamber and add the necessary parts to build this up into a separate hive.

As usual bees of flying age will return to the old site and will find their queen there in a brood chamber that contains very little brood and very few bees of nursing age; moreover there is a great deal of wax-making work to be done owing to the frames of foundation with which the brood chamber has been filled. Except for a small amount of brood on the frame containing the queen, conditions are now exactly the same as when the fourth step of the Pagden method has been carried out; the bees are deceived into believing that they have swarmed and make no further attempts.

The original brood chamber may be dealt with in any of the ways described under the Pagden method (page 76).

SWARM CONTROL

As you move among beekeepers you will hear much talk on the subject of swarm control. Ever since the introduction of moveable frame hives beekeepers have been looking for some simple, reliable and foolproof method of so manipulating or arranging their stocks that they will not swarm. Such a method if one could be found—would be applied to all stocks before they showed any signs of swarming and would prevent swarms throughout the season.

This idea is the Philosopher's Stone for which beekeepers seek in vain. No method has been devised which is at once simple and reliable; several are fairly reliable but are so complicated that it is usually as easy to cope with swarms before or after emergence as it is to operate the method of swarm control; several methods are simple but unreliable. It would be better to wait until you have gained experience before experimenting with swarm control but we shall end this chapter with a summary of the situation and a description of one important method frequently referred to in beekeeping literature.

Methods of swarm control fall into two classes:

(*a*) Baffling the bees' attempts to swarm.
(*b*) Making the bees imagine that they have swarmed.

Under (*a*) come such operations as cutting out queen cells every 9–10 days which has already been commented on; cutting off half of one of the queen's front wings so that she cannot join the swarm; standing the hive on a queen excluder so that the queen cannot come out when the swarm takes place; dequeening the hive during all or part of the critical swarming period. All these methods are simple but unreliable and lead to discontent and disorganization within the hive.

Under (*b*) comes the method of artificial swarming already described on page 78. All these operations may be carried out on a stock which shows no signs of building queen cells. When the original brood chamber has been set aside and the bees within realize that they are queenless they will at once set to and build queen cells. Conditions are then exactly as though one had discovered queen cells in a stock and had artificially swarmed it.

Successful artificial swarming requires two or three days of fine weather immediately after the operation has been carried out so that bees can fly freely and thus return to the old site. It also requires two complete hives for every stock of bees treated.

The famous *Demaree system* might almost be described as an artificial swarm within the hive itself; it is best carried out on a stock which has a well filled single brood chamber and is carrying at least one super.

Figure 32 illustrates the sequence of operations.

First, set aside the hive on a temporary stance.

Second, on the old site place a brood chamber filled with frames of foundation but having a space for the insertion of one more comb.

Third, examine the original brood chamber, discover the comb upon which the queen is standing and place this, together with all the attendant bees in the vacant space, in the new brood chamber.

Fourth, rebuild the hive as follows—floorboard, new brood chamber (and queen), queen excluder, super(s), queen excluder (optional), old brood chamber, quilt or crown-board.

As the bees sort themselves out the young nurse bees migrate upwards towards the brood in the old brood chamber at the top of the hive; the old foraging bees migrate downwards and occupy the lower parts of the hive; the queen, in the new brood chamber, finds herself surrounded by older bees, observes an almost complete lack of brood and is forced to check her rate of egg-laying owing to the absence of completed cells. The workers have a great deal of wax-making work to carry out and these conditions seem to satisfy the bees' swarming instincts.

The bees in the top brood chamber will probably raise queen cells and it is better to remove these about six days after the date of the operation. As the brood hatches out in the top chamber the bees will start to use this chamber as a super and to store honey in it so it may well be left in position until the end of the season. Workers which emerge from the brood in the top chamber will soon descend through the queen excluder(s) and join the queen below but a queen excluder is also a drone excluder and the drones are trapped in the top chamber. Lift the cover of the hive and set them free from time to time otherwise they will wedge themselves in the slots of the excluder in an attempt to escape and perish miserably; they may even block up the excluder completely.

If the Demaree method is applied to a hive at the beginning of a

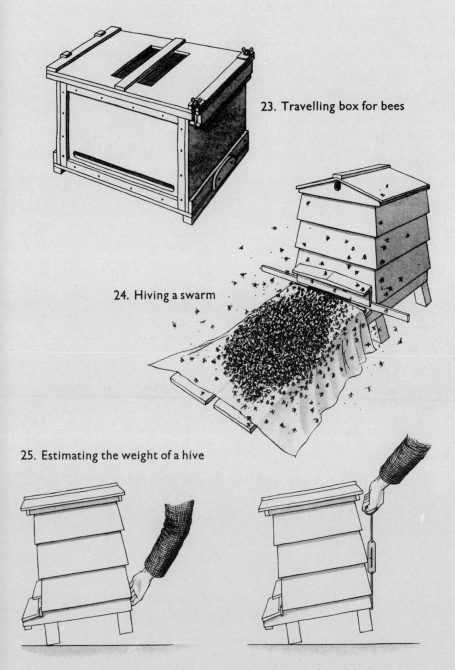

23. Travelling box for bees

24. Hiving a swarm

25. Estimating the weight of a hive

If a spring balance is used, first at the back and then at the front of the hive, the sum of the two readings gives the true weight of the hive. It is generally sufficient to feel the weight by hand

26. Examining a double brood chamber for queen cells

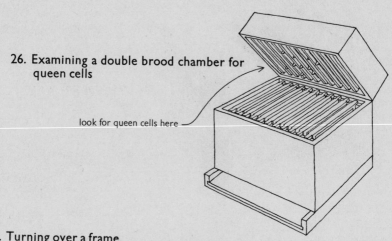

look for queen cells here

27. Turning over a frame

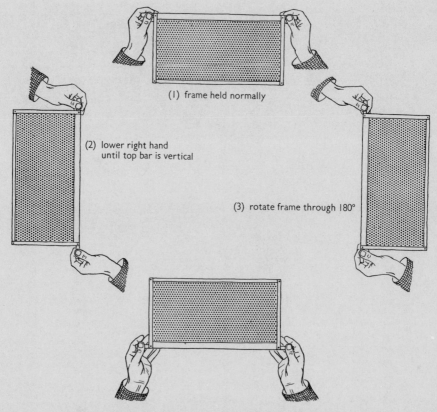

(1) frame held normally

(2) lower right hand until top bar is vertical

(3) rotate frame through 180°

(4) raise right hand so that top bar is again horizontal

spell of fine weather it is remarkably effective and may increase the honey crop in addition to preventing swarming. If the weather turns bad immediately after the method has been applied the bees tend to move upwards into the old brood chamber leaving the queen cold, hungry and nearly alone below the lower queen excluder. The Demaree method is about as reliable as the British climate; no more, no less.

Suggestions for further reading

Snelgrove, L. E. (1972) *Swarming: Its Control and Prevention.* (Published privately by Miss I. Snelgrove, Bleadon, Som.)

CHAPTER NINE

SPECIAL MANIPULATIONS

Operations carried out on a colony of bees are generally known as *manipulations*. Beekeepers even speak of manipulating a hive, a phrase which to my ear seems to have sinister undertones, as though the operations in question were not quite honest. Manipulations are best kept to a minimum. The less you disturb and disorganize your bees, the better they will work and the greater will your honey crop be. The simple routine manipulations necessary through the season have already been described in Chapter 7. The manipulations necessary in connexion with swarming have been described in Chapter 8. Now follow a few more manipulations which will be required from time to time when special circumstances arise.

UNITING BEES

One occasion when it is necessary to unite bees has already been met with when dealing with swarms by the Pagden method (page 76). Other occasions arise when a colony is found that has long been queenless and it is often better to unite it to a strong colony rather than to attempt the difficult task of requeening; when a colony is too weak in September to survive the winter (covering less than five frames) so that it must be united to a stronger colony; when a beekeeper has more colonies than he requires and wishes to reduce the number by uniting two or more of them together.

Bees do not unite readily and fighting with heavy losses will take place if proper precautions are not observed. This hostility is often said to be the result of each colony having an individual smell shared by all its inhabitants and thus being able to detect immediately the presence of intruders. Nobody, however, had ever brought forward clear evidence in favour of this theory and against it must be set the fact that a worker bearing a load of nectar is welcome in any hive as also are drones. Very probably the attitude of the intruders is the biggest factor in their unfriendly reception. Fortunately, all these difficulties may be overcome by the use of a simple sheet of newspaper to divide the two brood chambers to be united; by the time the bees have bitten their way through they appear to have acquired the same scent and have made friends with one another.

If you have a weak or queenless colony to unite to a strong colony, move the weak colony alongside the strong colony and leave it there for a couple of days until the bees are accustomed to their new situation. If either colony is short of stores, feed a pint or two of thick syrup to put the bees in a contented frame of mind. In the evening, when these operations are complete and the bees have ceased flying, remove the cover of the strong stock with the minimum of disturbance and spread a sheet of newspaper over the top of the hive. Prick the newspaper in half a dozen places with a pin or a sharpened match-stick because the bees' jaws, although excellent for nibbling or teasing away the paper, are not very suitable for making the initial hole. Then very gently place the weak stock on top, cover up and leave for 24 hours. Next morning shreds of paper will be seen at the entrance and on the alighting board and this is a sign that the uniting has taken place successfully and peacefully. The hive may now be opened and the combs rearranged as desired; usually the object is to reduce the combined colonies to a single brood chamber and to this end the frames in the upper chamber containing brood are put into the lower chamber or given to other stocks in the apiary while the bees from the remaining frames are shaken in.

If the lower colony is carrying a super this will not affect the operation. The super is left in place and the newspaper placed over it, but the need for uniting generally arises at the beginning or end of the season when there are no supers in position.

MAKING INCREASE

A beekeeper may wish to raise new stocks for several reasons. Perhaps his experience is now great enough to enable him to run a larger number of colonies; he may have to replace winter losses or losses due to disease; or he may wish to produce a stock for a friend to set his feet in the happy way of beekeeping.

In connexion with swarming we have already come across some methods of increasing stocks.

(1) Whenever a swarm is hived on a new site (page 75) this will build up and form a new colony.

(2) Whenever a colony and swarm are treated by the Pagden method the original brood chamber will produce a new queen and build up into a new colony (page 76).

(3) Whenever a colony shows signs of queen cells and is treated by the method of artificial swarming a new colony results (page 78).

All the above methods depend on a colony swarming or making preparations to swarm, but a beekeeper may desire to increase when none of his stocks shows signs of swarming. In such instances one of the following methods may be used. They all depend upon the fact that when bees are suddenly deprived of their queen they will choose larvae not more than 24 hours old or eggs, tear down the surrounding worker cells and build queen cells to accommodate the chosen eggs or larvae. These, when fed appropriately, develop into queens for you will remember that there is no difference between the egg from which a worker develops and the egg from which a queen develops; differences between the mature insects depend upon the different manner of feeding the larvae.

(4) A stock may be artificially swarmed even though it shows no signs of swarming as explained under the heading of swarm control (page 79). The queenless brood chamber which has been set aside will raise queen cells and soon become established as a new colony.

(5) If you already have two strong stocks it is an excellent idea to take from the first of these two frames containing eggs and young brood and two frames containing food; put these into a new brood chamber having first knocked off all the adhering bees. Be careful to avoid the frame upon which the queen is standing for she should never be knocked off a comb. Fill up the brood chamber with drawn out combs and, having moved the second hive to a new site, place the new brood chamber on the old site. Flying bees from the second hive will enter the new brood chamber, take care of the brood and raise the necessary queen cells. Fine bee-flying weather is necessary.

We have made a new stock by robbing stock No. 1 of brood and stock No. 2 of flying bees; if they are strong the losses will be made good in a very short time while the new stock will rapidly become established.

(6) One of the commonest, easiest and most generally used methods is to take about four frames together with the adhering bees from a strong stock and set them aside in a small hive to raise a queen of their own. For this method to be successful the following precautions must be taken.

SPECIAL MANIPULATIONS

(a) Take from 3 to 5 combs to form the nucleus. Two of the combs should be well filled with stores. One or two combs should contain sealed and emerging brood. One comb should contain unsealed brood and eggs.

(b) Make sure that the queen is not present on any of the combs used.

(c) Place the chosen combs in a small nucleus hive with the combs containing food outermost. A full sized brood chamber may be used if its size is reduced by a moveable partition known as a *division board*.

(d) Shake into the nucleus hive the bees from three additional combs. All bees of flying age will return to the colony from which the frames were taken so that, although a nucleus appears crammed full of bees when newly formed, the population will look quite thin on the following day and it is to compensate for this that extra bees are shaken in. Once again make sure that the queen is not among the bees shaken into the nucleus.

(e) Since a nucleus is thus denuded of its flying and foraging bees it must be fed liberally.

The disadvantage of a nucleus is that queen cells are built and queen grubs fed in a very small colony and there may not be enough nurse bees to feed the queen larvae adequately. If a frame carrying a sealed queen cell is available (from a colony that has swarmed) a nucleus is much more satisfactory and in this case a three-frame nucleus (two frames food, one frame brood) is adequate without the addition of extra shaken bees.

RAISING QUEENS

Since an old queen is of little use to a beekeeper and since queens sometimes disappear and stocks become queenless for no obvious reason, new queens are required from time to time as replacements. It is possible to buy all one's requirements from commercial queen breeders but this is an expensive procedure and is best resorted to when it is desired to introduce a new strain into the apiary owing to bad-temper, swarming propensities or other undesirable characteristics of the existing stock.

Stocks that have become queenless may be given a comb containing eggs and young larvae from another stock and, provided they have not been queenless too long and still have sealed worker brood, they will raise a queen for themselves.

Many of the methods already described under the headings of 'Swarming', 'Swarm Control' and 'Increasing Stocks' are also methods of raising queens but it will be convenient to summarize them here.

(1) Whenever a stock swarms and the swarm is hived on a new site the parent colony requeens itself and, since a number of queen cells are left behind, a nucleus may be made up as described on page 85 and a frame containing a sealed queen cell inserted. This will give a second queen. It is not necessary that the combs and bees forming the nucleus come from the same hive as the queen cells and if you have several hives and can therefore make up several nuclei without unduly weakening any hive several queens may be raised from the cells of one swarmed stock.

It is useless to insert two queen cells into a nucleus in the hope of raising two queens because the virgin queens, when they emerge, will fight until only one survives. Only one queen at a time may be obtained from one nucleus but the same nucleus may be used twice in succession if its strength is maintained. The virgin queen must be kept in the nucleus until she has started to lay and preferably until her worker brood has been sealed so that the beekeeper is sure that she is satisfactorily mated.

(2) Whenever a stock which has swarmed is treated by the Pagden method, the parent brood chamber, contains queen cells. This brood chamber may conveniently be split into two or three nuclei each containing a queen cell (page 76).

(3) Whenever the method of artificial swarming is applied to a stock the original brood chamber contains or will develop queen cells. When these are sealed, this brood chamber may be broken up into two or three nuclei each containing a queen cell (page 78).

(4) Whenever the Demaree method of swarm control is applied to a stock, queen cells will be raised in the top brood chamber (page 80).

(5) A beekeeper may own a stock of bees which shows desirable characteristics such as good temper, hard working and little tendency to swarm and may wish to raise queens from this stock. If such a stock shows no tendency to raise queen cells it may be provoked into doing so by removing the frame containing the queen and housing her temporarily in a small three-frame nucleus. The main stock, finding itself queenless, will raise queen cells and these may be put into nuclei as desired. Any excess queen cells are destroyed

and the queen reunited to the colony on the 9th or 10th day after her removal.

The paternal aspect of queen rearing we have not so far mentioned, but it is obvious that the drone plays his part in the character of the progeny of the mated queen. Over drones one has very little control but it is worth remembering that a queen most often mates with a drone from the hive in which she was reared. Where a beekeeper is trying to raise further queens of a satisfactory strain it is better to use method (3) since drones of the same colony are more likely to be present in the queen rearing nuclei.

(6) In all the above methods queen cells have been raised and sealed in full strength stocks capable of adequately feeding the queen larvae, and if the finest quality queens are to be obtained this is essential. If however, a strong nucleus is made up and contains larvae less than 24 hours old, and eggs, it will raise a queen for itself. This method is not recommended except in an emergency.

QUEEN INTRODUCING

It seems almost superfluous to say that it is no use attempting to introduce a queen to a colony that already has one but, since I have made that mistake myself more than once, let us set this down as the first requirement for success. Nor must the colony have queen cells. **Removing the old Queen.** If an old or unsatisfactory queen has been found and removed, allow about two hours to elapse so that every bee in the hive has time to become aware of the queenless state of the colony before attempting to introduce the new queen. After about 24 hours they will have started to build queen cells and are no longer in a receptive frame of mind; if this happens one must wait one week and then destroy all queen cells formed. No more cells can then be started since all eggs will have hatched and all larvae will be too old; the bees are now powerless to raise a queen of their own and are once again prepared to receive an offered queen. **Conditions for Introducing the Queen.** The colony to which the queen is to be introduced must be prosperous and well supplied with food; if not, feed for two days previously. The colony must contain young bees for attendance on the newly-introduced queen. In colonies that have been long queenless there may not be any bees of 1–6 days old and in such cases one or two frames of emerging brood should be inserted a day or two previously.

The Newspaper Method. If you have raised your own queen and she is mated and laying in her nucleus, the easiest, simplest and safest method of introducing her is to unite the colony and nucleus by the newspaper method. If you have purchased a queen or if you wish to use your nucleus a second time then you will have to introduce your queen as an individual separated from her brood and eggs. This is a more difficult operation because the sudden appearance of a laying queen in a colony that knows itself to be queenless is a miracle to the bees and she is viewed by them with suspicion and mistrust. The queen, finding herself in strange surroundings, adopts a truculent and hostile attitude. Lastly the queen does not have the hive odour.

Simmin's Direct Method. One of the simplest methods of introducing a queen is that known as Simmin's direct method. The hive is dequeened (if necessary) in the afternoon and allowed to stand for two or three hours. Half-an-hour before dusk the new queen is placed in a matchbox, alone and without food, and the closed matchbox is placed in the waistcoat pocket or other suitable warm place. When half-an-hour has elapsed and the bees in the hive to be requeened have ceased flying, go to the hive and remove the cover of the feed hole as quietly as possible, take the matchbox and slide it open about $\frac{1}{4}$ inch, placing it opening downwards over the feedhole. Cover up with a cloth and do not disturb the hive for a week. A modification of this method is to pour lukewarm water into the matchbox (without waiting half-an-hour) and then drop the wet queen directly into the hive.

The period of waiting or the wetting both appear to diminish the queen's special odour, while the fact that she is very hungry in the first case, or wet and miserable in the second case, removes her truculent attitude so that she is readily accepted by the bees.

Introducing a Purchased Queen. If a queen is purchased she will arrive by post in a *cage* which serves at once as a travelling box and a means of introduction. Such a cage is shown in Figure 33. A block of wood about 4 inches × $1\frac{1}{2}$ inch × 1 inch is partly bored out to form two compartments, one larger than the other. In the larger compartment is placed the queen with about a dozen attendants. The smaller compartment is filled with a stiff paste made by mixing liquid honey and icing sugar and this serves the bees for food during the journey since a hole is drilled between the two compartments to give the bees access to the candy. A piece of wire gauze is tacked over the open side of the block and this forms a cage and retains the bees. At each end of the cage is bored a round hole about ⅜ inch diameter, one giving access to the queen's compartment and one to the candy compartment but both are covered with gauze before despatch.

When the cage is used for introduction, the cover from the small hole giving access to the candy compartment is removed and the cage is set on top of the brood chamber frames, at right angles to them and with the wire gauze vertical (Figure 34). The bees in the hive now begin to eat away the candy while the bees in the cage are still eating their way into it from the other side. In about 48 hours the two parties meet and the queen leaves the cage through the passage so formed. During this period the queen will have acquired the hive odour and the hive bees will have become accustomed to her presence even though she has been behind bars which protected her from their initial hostility. The queen, glad to be free, displays no animosity and the hive bees do not regard her as a stranger; unmolested she begins to lay. Once again do not disturb the stock for 7 days. A queen is not really safe until she has begun to lay and some of her eggs are hatched. A purchased queen may be taken from her cage and introduced by one of the direct methods if desired.

FINDING THE QUEEN

Several beekeeping operations involve finding the queen and many beginners find this difficult. Here are some suggestions which may help. When opening a hive, preliminary to looking for the queen, use a minimum of smoke and disturb the bees as little as possible. The queen will be found on a frame containing eggs and empty cells; rarely on frames full of sealed brood or on the end frames of a brood chamber. The queen tends to avoid light so, when removing a frame from the hive, examine the unexposed face first. Normally the queen is surrounded by a circle of twelve or more attendants who stand round her facing inwards but when smoke is applied to the hive the attendants usually disperse and the circle may not be seen. The queen moves across the comb in a slower and more dignified fashion than the workers, she stands a little higher, her colour is usually a little different, and, of course, she is larger.

If all other methods fail, the brood chamber may be put on a new site leaving the supers on the original site. In an hour or two, if the weather is fine, all the flying bees will be in the supers and the queen will be found easily in the half-empty brood chamber.

MOVING BEES

In winter, when no bees are flying, hives may be moved as required but in summer, owing to the bees' habit of returning to a site and not to a hive, the problem presents some difficulties. If bees have to be moved more than about $1\frac{1}{2}$ miles, that is outside their normal flying range, no difficulties are encountered. The bees, finding themselves in a new location, spend a little time in committing the new district to memory and then return on all future occasions to the new site.

If hives have to be moved within the apiary during the summer they can be moved about a yard a day and, although even this causes confusion, the bees can refind their home. If this technique proves too tedious, the hive may be moved into its new situation and the entrance blocked up with a handful of grass so that the bees have to bite and struggle to emerge. This draws their attention to the fact that there is something different about their situation and they will accordingly mark their new location.

Suggestions for further reading

Snelgrove, L. E. (1981) *Queen Rearing.*—Published privately by Miss I. Snelgrove, Bleadon, Som.

Cook, V. (1986) *Queen Rearing Simplified.* BBJ.

CHAPTER TEN

THE HARVEST

'To travel hopefully', said Stevenson, 'is a better thing than to arrive'. The honey harvest is the point towards which the beekeeper has been travelling throughout the long summer, the climax of the beekeeping year, yet I must confess I prefer looking after the bees themselves to the sticky and laborious process of extracting their honey. Some beekeepers persuade their wives to accept the responsibility for this part of the year's work and there is a good deal to be said for this idea.

The operations involved in working up the honey crop are:

(1) Sorting out the combs so that those that are largely unsealed or otherwise unsatisfactory are returned to the bees.

(2) Cutting off the cappings of the honey cells with a sharp knife, generally known as *uncapping*.

(3) Spinning round the uncapped combs in a rapidly rotating cage so that the honey is flung out of the cells, generally known as *extracting*.

(4) Straining the extracted honey to remove fragments of wax, pollen, etc.

(5) Bottling the strained honey.

(6) Rendering down the cut-off cappings into beeswax.

PREPARATION

Before we look at the above operations in detail we must collect together the necessary apparatus. This is what will be required.

A Working Place. Most beekeepers who own only a few hives use the kitchen and if your wife's assistance has already been enlisted the necessary permission should be readily forthcoming. A kitchen is a convenient place since hot and cold water are available for washing sticky hands and implements from time to time. If you have to work in some other place have a large damp towel and some damp rags available. The room should be proof against bees and wasps since they are attracted by the strong sweet scent of newly extracted honey. The room should be warm, especially if the honey combs have been kept for a little while before extraction; cold honey is difficult to extract. Spread some newspaper on the floor and wear a clean overall.

Uncapping knife. A heavy knife heated by dipping into hot water was once commonly employed but, provided it is very sharp, any knife may be used cold and this is far more convenient. The ordinary domestic carving knife or bread knife with a plain blade both serve excellently but I have not found bread knives with saw-tooth edges very satisfactory because they seem to tear the comb rather than to cut through it, and they cannot easily be sharpened. The best knife for uncapping is known as the 'Granton' (Figure 35) which has a dimpled pattern on both sides of the cutting edge and is easily sharpened.

Cappings Tray. Some receptacle must be provided to catch the cut-off cappings and accompanying honey. A large meat dish will serve, but a clean enamel bucket is better since you can fix a bar of wood across the top to support the frame while uncapping (Figure 38).

Extractor. This consists of a rotating cage into which the uncapped combs are placed. A handle and the necessary gearing are provided so that the cage may be rotated at speeds up to about 200 r.p.m. The cage is contained in a metal cylinder or tank open at the top and provided with a tap at the bottom. The outflung honey runs down the inside of the cylinder and accumulates at the bottom.

Figure 36 (*a*) shows the radial type of extractor in which the uncapped combs are arranged like the spokes of a wheel and the honey comes out from both sides of the frame simultaneously. The cage will hold from 12 to 20 frames at one loading but extraction may take from 5 to 15 minutes so that this type of machine is usually driven by an electric motor and is more suitable for the large beekeeper.

Figure 36 (*b*) shows the tangential type of extractor in which the uncapped combs are arranged with one side facing outwards. This machine extracts honey only from the outward faces of the combs and it is necessary to stop and reverse the combs half-way through the process. The usual capacity is four combs at a time and the machine works quickly, extracting one set of sides in about two minutes.

An extractor is an expensive piece of equipment and **the** four-frame tangential type will cost about £36 (1975). Most Beekeepers' Associations have an extractor which they will hire out to members for a modest fee, but if you wish to buy your own you will find that the tangential type capable of taking four deep frames will meet all your needs until you own 10 or more colonies.

Strainer and Bottling Tank. This is a cylindrical tank with a tap at the bottom rather similar to the extractor but without the rotating cage, and rather smaller. The top portion is removable and has a perforated metal strainer in the base. Additional straining is obtained by tying a piece of muslin over a flange in the base (Figure 37).

Containers. Honey is usually put into glass jars so that its attractive appearance may be appreciated. The old-fashioned tall jar, so designed that the bottom was beyond the reach of the average jam spoon, has now been largely replaced by the squat Ministry of Agriculture jar available in $\frac{1}{2}$ lb. and 1 lb. sizes at about 48p per dozen. For those who wish to store honey in bulk 7, 14 and 28 lb. lacquered tins are available with lever lids.

SORTING THE COMBS

It is important that extracted honey should consist only of ripened honey, from cells sealed by the bees; thin, unripe honey from unsealed cells makes your entire crop liable to fermentation if too great a quantity is included. All combs which consist largely of unsealed cells of honey are set aside and either extracted separately or given back to the bees. It is generally laid down in beekeeping textbooks that no unsealed honey should be included in the main crop but, in practice, beekeepers regard this as a counsel of perfection and actually any comb containing not more than one quarter of its cells unsealed may be safely included in the main crop.

Sometimes beekeepers go through their combs a second time, holding them up to the light and grading them into 'light' and 'dark' classes. The colour of honey produced at different times during the same season varies considerably but the second sorting is not usually carried out unless a beekeeper requires both light and dark honey for exhibition purposes.

UNCAPPING

This operation is shown in Figure 38 from which you will see that:

(*a*) The frame is placed in an almost vertical position with its bottom lug resting in a hole made in the wooden supporting bar.

(*b*) The frame is held sloping slightly outwards and away from the body so that the cut off cappings fall away from the face of the comb into the bucket.

(*c*) Cappings are cut off from the side away from the operator and the cut is made upwards from the bottom of the comb.

(d) The cut should be between $\frac{1}{16}$ inch and $\frac{1}{8}$ inch below the surface of the comb but since the comb surface is often uneven one is frequently compelled to cut deeper. Do not be disturbed if a good deal of honey seems to go into the receptacle along with the cappings; it is not wasted because the cappings are drained and the drained honey is added to the main bulk.

(e) There will be low places in a comb over which the knife passes without uncapping the cells. These small areas will have to be uncapped individually using the tip of the knife.

There is considerable skill in the process of uncapping so do not be distressed if your first efforts seem slow and unsatisfactory. With practice you will soon be able to remove the cappings all in one piece from one side of a good comb. Wash both your hands and the knife frequently, because this is the stickiest part of the proceedings. See that no particles of wax are sticking to the knife blade, and sharpen the knife from time to time.

EXTRACTION

After uncapping, the frames should go straight into the extractor. In a radial machine they are placed with the top bars outward, i.e. away from the spindle. In a tangential machine (which you are most probably using) the top bars may be either leading or trailing but see that the outer face of the comb is in contact with and supported by the wire grid of the rotating cage. In either case place combs of approximately equal weight on opposite sides of the cage so that it is balanced. Begin turning very slowly. With the radial extractor the speed is slowly increased until no more honey is flung out of the combs. In using the tangential extractor it is best to extract partly the first face of the combs keeping the speed of the extractor low, reverse them and extract the second side again starting slowly but rising to full speed in 1–2 minutes then once again reverse the combs and complete the extraction of the first face at full speed. The droplets of honey are flung out of the cells, strike the inner walls of the extractor and run down to the bottom where the honey is run off from time to time through the large-bore tap provided.

Extraction is much quicker, easier and more complete if the honey is extracted while it is still warm from the hive. If it is not possible to extract immediately after removal of the supers, keep them in a very warm place for several hours before extraction.

STRAINING

Honey run off from the extractor contains particles of wax, pollen grains and minute air bubbles and the object of the strainer and bottling tank is to remove all these as completely as possible.

Before use, tie a piece of muslin or bolting cloth over the bottom of the straining section of the bottling tank and then run in honey from the extractor. The perforated metal retains all the larger fragments of wax while the muslin removes finer particles and perhaps some pollen, although pollen grains are always present even in the most carefully strained honey. After a time the muslin becomes clogged and the straining becomes tediously slow. When this happens it should be replaced by a second piece while the first piece is washed and dried.

After straining, the honey must remain in the bottling tank for 24 to 48 hours so as to give time for the minute suspended air bubbles to rise to the surface where they form a white froth or scum. There is no harm in either the air bubbles or the froth but they both detract from the clarity and attractiveness of the bottled honey.

When the straining of the main bulk of the honey is complete the cappings may be put into the strainer and the honey from them allowed to strain into the bottling tank; or the cappings may be suspended in a muslin bag and allowed to drain. As with extraction, the process of straining is made much easier and quicker if the honey is warm.

The very small scale beekeeper may, if he wishes to avoid the purchase of a bottling tank, allow his honey to stand for 24 hours in an extractor and then strain it directly into bottles, but the process is painfully slow and the clarity of the honey is inferior.

BOTTLING

Remember that English honey is a connoisseurs' product and take every care to bottle it so that it looks as attractive as it tastes. Use standard jars of a uniform size and not odd jam jars. See that the jars are washed and spotlessly clean and are quite dry before use. See that the lids are bright and free from rust and that each contains a new and dust-free waxed cardboard wad.

In bottling let the stream of honey fall on the sloped side of the jar so avoiding trapping additional air bubbles. For the same reason do not let the honey drop into the jar from too great a height.

Labelling. The appearance of your honey is much improved by a label. All appliance dealers offer a range of designs and some Beekeepers' Associations issue a label for the exclusive use of their members. The design of many labels is singularly lacking in taste so choose carefully. Do not let your label be so large that it prevents a prospective purchaser from viewing the contents of the jar and appreciating the appearance of your honey. Do not choose a gaudily coloured label which will distract attention from the honey itself; black and gold always looks restrained yet attractive. If you intend to sell some of your honey it is a legal requirement that both the name and address of the producer and the nett weight of the contents appear on the jar. Prepare and bottle your honey in such a way that you are proud to see your name on the jar.

GRANULATION

Honey when freshly bottled is clear, but on standing, and as the weather becomes colder, it begins to crystallize or granulate and usually becomes solid although the hardness is variable (see page 17). The onset of granulation may be delayed by heating the honey for a few hours at 120° F. The best way of doing this is to immerse the jars in water at this temperature and to maintain it with a very small flame. If this temperature is exceeded both the aroma and colour of the honey may be spoiled. Keep the lids of the jars tightly screwed down during the process to retain the aroma; there is no danger of the jars bursting. If it is required to change granulated honey back to clear honey apply the same heating process when the crystals will redissolve and the honey become clear again. Honey that is half clear and half granulated looks extremely unattractive and should never be offered for sale. Turn it into clear honey by heating.

Most small beekeepers who want some granulated honey allow it to crystallize in the jar and this practice is very convenient but has two disadvantages. In the first place the honey may set too hard so that you may even bend the spoon when getting it out of the jar. In the second place *frosting* (a white appearance largely due to slight shrinkage of the honey during granulation) may take place and, although this is no reflection on the quality of the honey, the appearance is spoiled. Both these defects may be overcome by allowing

28. Knocking bees off a frame

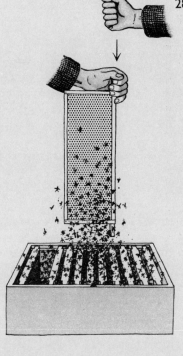

29. 'Rapid' feeder

GENERAL VIEW

CROSS SECTION

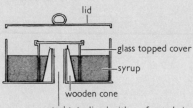

lid
glass topped cover
syrup
wooden cone
central tube lined with perforated zinc

30. Clearer board with Porter bee escape

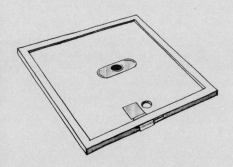

detail of Porter bee escape

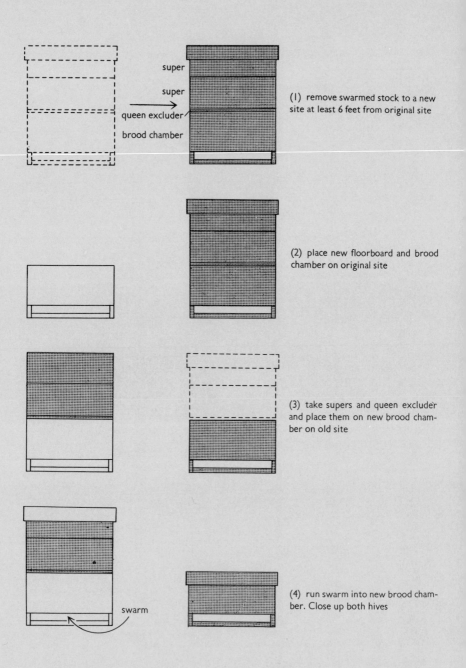

31. The Pagden method of dealing with swarm and parent colony

THE HARVEST 97

honey to granulate in bulk in a 14 lb. tin. It should then be kept at 110° F. for an hour or two until it is partly melted and, on stirring, a creamy mixture of crystals and liquid honey is obtained. If this is then bottled it will never set absolutely hard, nor will frosting occur. Light coloured honey is most attractive on granulation and clover honey is pre-eminent in this respect.

WAX

The drained cappings are still very sticky with honey and the best way to purify them is to melt them with about an equal volume of rainwater, or tap water acidified with vinegar (1 teaspoonful per pint), in an enamel bowl. After cooling, the cake of wax can be removed from the surface of the water but much dross will remain on its underside. Scrape off the loose dross and repeat the melting process until the water is no longer coloured. The resulting cake of wax will still contain dross and will require melting and straining to remove the last traces of impurity. The cake of wax may be suspended in the oven in a muslin bag and allowed to melt and drip into a basin below it, or it may be tied in a muslin bag with a stone to act as a sinker, placed in a large vessel of rainwater and brought to the boil. The wax will melt, pass through the muslin and rise to the surface of the water leaving the dross behind. This last method may be used also for the recovery of wax from old and damaged combs. Various devices known as wax extractors are on the market and they certainly make the process cleaner and quicker if the amount of wax to be rendered justifies their purchase.

CLEARING UP

Your extractor and settling tank are probably made of tinplate and must therefore be very carefully washed and dried before putting away. Warm water is a help, but do not let it be hotter than you can comfortably bear your hand in otherwise you will melt the fragments of comb and they will stick to your apparatus and clog your strainer.

Next there is a pile of empty super combs from which you have extracted your honey, to be dealt with. One method is to replace these in the supers just as they are, sticky with honey. Wrap up each super in newspaper and store in a dry cool place. This method is easy, the sticky combs discourage the wax moth and, when the supers are put into use again, the bees are attracted to them by the traces of honey remaining. Alternatively, the supers of wet combs may be placed on the hive over the clearer board when, on opening the special slide provided, the bees will re-enter the supers, clean up the combs and take down into the brood chamber all remaining traces of honey. In 48 hours the slide may be closed when the bee escape will come into operation and the supers will once again be free from bees in a further 24 hours. This method is decidedly cleaner and neater and makes for easier winter storage of the empty supers.

In addition you have a number of combs containing unsealed honey which it is intended to return to the bees. If your brood chamber includes a box of shallow frames you will probably be able to find some empty ones in early September and to replace these by your unsealed combs of honey. Otherwise put all your unsealed combs in a super and place this beneath the brood chamber; the bees will carry up the honey into the brood chamber and the empty super may be removed in late October or in the following spring.

If you decide to extract the unsealed honey, dilute it with an equal volume of water before feeding it back to the bees. Honey feeding causes great excitement and may provoke outbursts of robbing; it should be carried out preferably in late autumn or early spring.

EXHIBITING

Like gardeners, beekeepers love to vie with one another by exhibiting their finest products and to this end all Beekeepers' Associations organize honey shows and appoint honey judges.

In order to increase the number of classes and to make judging easier, clear honey is subdivided by colour into three groups, light, medium and dark; there is no clear cut division between these groups and the limits are defined by the aid of standard coloured pieces of glass. There are also classes for granulated honey, beeswax and products such as honey cakes and mead.

There are always a number of classes for beginners (usually defined as those who have never won a prize at a honey show), classes restricted to members of the Association organizing the show and, in a large show, classes open to beekeepers throughout the country. Generally two or three jars of honey are required for an entry. You should certainly enter the honey show of your Association. Here are the main points which the judges look for and to which you, as an exhibitor, must pay attention.

Cleanliness. Your clear honey must be quite free from visible dirt in suspension or any bubbles or scum floating on the surface. Strain your honey two or more times.

Clarity. The clearer and brighter your honey looks, the better chance it has of winning. Haze, if the honey has been well strained, is the result of suspended air bubbles so let your honey stand for two or three days at 90°–100° F. before bottling, thus giving the bubbles a good chance to rise to the surface.

Density. The thicker your honey is the better, so do not include any honey from unsealed cells in honey intended for exhibition.

Colour. Check your exhibit against the standard coloured glasses and see that it is in the right colour class otherwise you may be disqualified on purely technical grounds. See also that all the jars of honey which constitute one exhibit are all exactly the same colour.

Flavour and aroma. When a judge has eliminated a large percentage of entries for lack of the qualities set out above, he will make his final decision by actually smelling and tasting the honey. If your honey is not of good flavour to start with there is nothing you can do about it, but you can avoid spoiling a good flavour by excessive or prolonged heating or by undue exposure to the atmosphere.

General Appearance. See that your honey jars are clean and free from defects; make sure that they contain at least the statutory pound of honey; verify that, if any particular type of jar is specified, you are using it; take care that the lids are clean, bright and unscratched and that the waxed wad is clean and new. All exhibits are anonymous and no label may appear on the jar except one small label carrying an identifying number and provided by the show secretary.

Suggestions for further reading

Anon. (1953) *Honey from Hive to Market.* Ministry of Agriculture Bulletin No. 134. O.P.

Herrod-Hempsall, W. (1948) *Producing, Preparing, Exhibiting and Judging Bee Produce.* British Bee Journal.

CHAPTER ELEVEN

TROUBLES

Except for swarming we have only hinted at the other troubles which may beset the beekeeper. Now, in a final chapter, we must face these difficulties.

These troubles may be divided in three groups (*a*) diseases, (*b*) pests and enemies, (*c*) disorganization, and under these three headings it will be simplest to give a section to each trouble, describing in each the signs, causes and treatment.

DISEASES

Since the bee exists in two different states, the grub or larva and the perfect insect at maturity, so there are two sets of diseases which attack the bee. The brood diseases attack only larvae and no adult bee can ever be infected by them; adult bees live healthily (if not happily) in a hive containing larvae suffering from brood disease. The adult bee diseases never attack larvae and these develop normally into mature insects even when the adult population of the hive is diseased.

Brood Diseases

American Foul Brood (or A.F.B.)

SIGNS. The larvae die after their cells have been sealed so the first thing to look for is abnormal cappings to brood cells. Cappings are slightly sunken since the enclosed larvae start to shrink under the influence of the disease. Cappings may be pierced due to the efforts of the adult bees to extract from the cell and eject from the hive the diseased larvae. If such cells are opened by the beekeeper the grub inside will have turned yellow or brown. A little later the grub begins to decompose and, if a matchstick be thrust into the cell and twisted round and then slowly withdrawn, the contents may be drawn into a brown slimy thread one or two inches long. In the latest stage of the disease the grub dries up to a tightly adher-

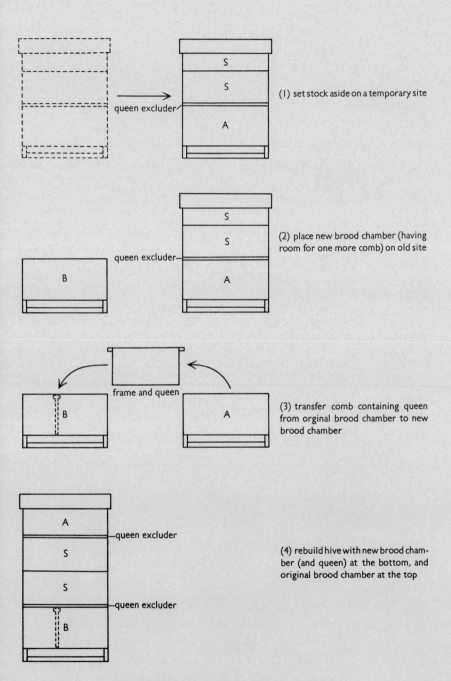

32. The Demaree method of swarm control

33. Postal queen cage

queen cage without bees, candy or wire gauze, to show arrangement of passages

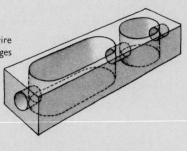

queen cage complete

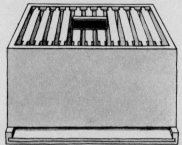

34. Queen cage in position

35. Uncapping knife 'Granton pattern'

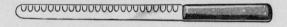

36. Honey extractors

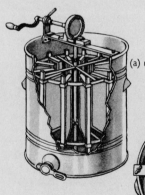

(a) radial extractor

tangential extractor (Taylor's 'Tubby' model)

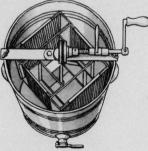

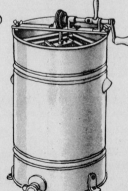

from above (lid supplied but not illustrated) and side view

ing scale which lies in the lowest angle of the cell (now completely uncapped by the bees). These signs are illustrated in Figure 39.

CAUSES. This disease is caused by bacteria and is therefore highly infectious. The spores from which these bacteria develop can lie dormant for years in honey; they can also survive for considerable periods in infected hives and equipment. The disease is spread from hive to hive by bees robbing honey from an infected colony; by the use of secondhand infected equipment; by the drifting of drones and young bees from one hive to another; by the beekeeper himself using contaminated hands or tools. Within the hive the disease is spread by nurse bees feeding larvae with contaminated honey.

TREATMENT. If any of the signs mentioned above are observed, you must at once advise your local bee expert. If he confirms your observations, he will advise the local Bees Officer (Min. of Agriculture) and he has authority, under the Foul Brood Diseases of Bees Order, 1967, to order their destruction and to be present when the infected colonies and appliances are destroyed.

There is no reliable remedy for this disease and the infected colony will have to be killed and then the bees, combs and quilt will have to be burnt and buried. This destruction is drastic and heartbreaking but is required by law as the surest method of checking the spread of the disease. After the destruction of the bees, all brood boxes and supers must be sterilized by scorching them with a painter's blow lamp until the wood turns light brown. Feeders, smokers, hive tools and queen excluders should be scrubbed with a warm, fresh solution of 1 lb. of washing soda plus $\frac{1}{2}$ lb. of bleaching powder in a gallon of water.

Prevention must be the watchword since cure is impossible. If you buy secondhand hives disinfect them thoroughly by scorching before putting them into use. Never buy secondhand combs. Never feed honey of unknown origin to your bees. Always ask for a guarantee of good health when buying bees. If you gather or are given a stray swarm of bees, hive it separately and do not unite it to any healthy stock until its queen has started laying and her sealed brood is seen to be healthy.

Whenever you open a hive keep a sharp lookout for signs of A.F.B. and, if you notice them, close your hive, restrict the entrance and send for expert advice at once.

European Foul Brood (or E.F.B.)

SIGNS. The larvae die before sealing, turning yellow and then very dark brown. The grub takes up unnatural attitudes and often uncoils itself. Sometimes the grub may be seen wriggling about in its cell. When the larva dies it forms a sticky granular mass which cannot be drawn out in a thread with a matchstick. When the dead larvae dry up the resulting scales are easily detached from the base of the cell and are usually removed by the bees. This disease is much more uncommon than A.F.B. but spreads rapidly through the apiary and is therefore more dangerous. The disease is usually accompanied by a foul smell resembling that of bad fish; sometimes the smell is sour and not so unpleasant but this sign is not always present. (Figure 40.)

CAUSES. This disease is caused by bacteria (different from the A.F.B. bacteria) which kill the grub when it is 4 to 5 days old. Other remarks as for A.F.B.

TREATMENT. This disease is also covered by the Foul Brood Diseases of Bees Order and you are bound by law to report it. No satisfactory cure is known and the bees and combs have to be destroyed. The procedure adopted is exactly the same as described under A.F.B.

Chilled Brood

SIGNS. The larvae usually go black (not brown) and die in large areas usually around the edges of the brood nest. Larvae die at all ages and in all stages of development.

CAUSES. Properly speaking this is not a disease at all but it is included because it is common and is often mistaken for one of the foul brood diseases. It is caused either by the beekeeper opening the hive at unsuitable times and exposing brood to the cold external air or by a sudden and unseasonable spell of cold weather forcing the bees to contract the brood nest and to abandon some of the developing brood.

TREATMENT. The bees will soon throw out the corpses of the dead and all signs of the trouble will disappear but the colony will be set back by the loss of several thousand potential bees. Feeding will encourage the queen to make good the loss and help in keeping the hive warm. Above all, do not open the hive on cold spring days.

Other known brood diseases include Chalk Brood, Stone Brood, Sac Brood and Addled Brood. These are neither so common nor so dangerous as those already described. Details may be found in the larger text books.

Adult Bee Diseases

Acarine

SIGNS. An excessive number of dead bees are to be seen in front of the hive. Very many bees are to be seen 'crawling', that is walking on the ground away from the hive in all directions, climbing up grasses or sometimes forming little groups huddled together as if for warmth. Often the hind wing sticks out sideways from the body at an awkward angle while the fore wing is folded back normally, an appearance known as *K-wing*.

CAUSES. This disease is due to a parasitic mite entering the front breathing tubes of young bees, living and breeding there and slowly blocking up the tubes with increasing population and accumulated excreta. These breathing tubes supply air to the wing muscles of the bee and when the oxygen supply is diminished the bee is unable to fly; it can only crawl. Also the mites live on the juices of the bee's body, thus further weakening it.

If any of the signs set out above are noticed then the disease may be confirmed by taking a dead or dying bee and pulling off its head with a pair of tweezers. By the aid of a simple lens the breathing tubes can be seen in the aperture and removal of the first segment or collar of the thorax makes the tubes more visible. The breathing tubes should be pure white; if they are black, brown or blotchy acarine disease is confirmed. (See Figure 41.)

The acarine mite can only live for a few hours outside the body of the bee and can only enter the breathing tubes of very young bees. The disease is spread by drones and young workers infested with the mite drifting into other hives. Sterilization of combs, hives and tools is not necessary since this is not a bacterial disease.

TREATMENT. The following mixture, known as *Frow Mixture* after its discoverer, gives off a vapour which appears to kill the mite without harming healthy bees.

Nitrobenzene 2 parts   Oil of Safrol 1 part   Petrol 2 parts

The vapour of this mixture is however, harmful to brood and the treatment must be applied only when there is no brood in the hive, usually in November or early February. The strong smell of the mixture seems to attract robber bees and entrances to hives should be restricted or even completely closed with perforated zinc during treatment. A piece of felt is placed over the feed hole and upon it is placed daily 30 minims ($\frac{1}{2}$ teaspoon) of the above mixture; this treatment is repeated daily for 6 days and the pad is left in position for a further week. The dose may be doubled and given at two-day intervals on three occasions if desired.

An alternative and very effective treatment is to use methyl salicylate (or oil of wintergreen). This treatment must be applied in the summer when the weather is warm. Obtain a flat tin not more than $\frac{3}{4}$ inch deep, pack it tightly with cotton wool, make a number of holes in the lid with a nail and pour about $\frac{1}{2}$ oz. methyl salicylate, obtainable from most chemists, on to the cotton wool. Place this tin on the floor of the hive underneath the brood frames and leave it there during the whole summer. Although methyl salicylate has a strong smell it does not upset the bees, harm the brood, provoke robbing, nor does it taint the honey.

Methyl salicylate does not appear to kill the acarine mite but it prevents it migrating from an infested bee to a healthy one. The infested bees will ultimately die, the mites will die with them and the stock will recover its health. Many beekeepers apply this treatment to all their stocks every year as a preventive measure.

A third method is the use of 'Folbex' strips which consist of thick paper impregnated with Chlorbenzilate. When a strip is lighted, it smoulders and releases a vapour which kills acarine mites but does not harm bees or brood; it can therefore be used at any time of the year.

An empty box is put under the brood chamber and the smouldering strip is placed on the floor-board. Repeat two or three times at weekly intervals.

Nosema

SIGNS. Bees crawl but do not walk far away from the hive. Dying bees fall over on their sides or backs with trembling movements of the wings and legs. Bees often seem to be suffering from dysentery and if the abdomen of a dead bee be squeezed the contents of the bowel shoot out. The colony becomes weak, often apparently without cause, particularly in the spring.

CAUSES. A very small parasite which lives in the stomach and intestines of the bee; in this the disease resembles malaria. The spores of the disease pass out of the bee with the contents of the bowels and, since the disease is often accompanied by dysentery, the combs be-

come infected and the disease is transmitted to healthy bees. The spread of the disease is usually slow but virulent outbreaks occur.
TREATMENT. A cure is now available for this disease known as 'Fumidil B'. This is added to syrup fed to an infected colony. Some beekeepers add this to their winter feed whether the colony is infected or not; this is a wise (but expensive) precaution. A strong colony will often recover due to its own house cleaning efforts. If the strength of the colony justifies it, the bees may be rehoused in a clean hive on frames of foundation and fed liberally.

The infected frames may be cleansed by subjecting them to the fumes of strong acetic acid for a week; after this they must be exposed to the air for at least a week before re-use.

Paralysis

SIGNS. Bees become hairless, shiny and dark coloured. Healthy bees may be seen ejecting sick bees from the hive, the latter protesting and struggling. An excessive number die often with tongues extended, with bowels distended and sometimes with trembling movements.

CAUSES. A number of diseases and troubles are gathered together under this heading including deficient queens, or poisoning.

TREATMENT. Paralysis often disappears of its own accord. If the trouble is poisoning it comes from fruit spraying or from poison pollen and as the cause is only temporary the signs die away though the colony may be much weakened. If the trouble does not disappear, requeening the stock will often put the matter right.

Poisoning by Insecticides

SIGNS AND CAUSES. Bees dying and dead with swollen abdomens. Unusual number of dead in front of the hive. If poisoned pollen has been brought in, larvae may be killed also. These signs are typical of the poisoning produced by arsenic; arsenic is a stomach poison and acts slowly so that bees get back to the hive before they die.

TREATMENT. This trouble is of course not a disease but arsenical poisoning may be mistaken for one. Spraying is now so frequent and widespread that the keeping of bees within flying range of a commercial orchard is usually a doubtful proposition. The spraying that goes on in private gardens is not sufficiently widespread to cause serious trouble but if you use insecticides yourself you will avoid damage to your colonies by observing the following precautions.

(1) Never spray (or dust) open blossoms. Spray while in bud—the 'pink bud' stage with apples—or after the petals have fallen. (2) If there are flowers such as dandelions, upon which the insecticide might fall when the trees are sprayed, cut them down. (3) Add 1% lime-sulphur to all sprays. The smell of this repels the bees. (4) See that the bees have a source of drinking water well away from the spraying area. (5) If you have a neighbouring fruit farmer see whether you can persuade him to adopt these simple precautions.

Of the generally used insecticides DDT seems little harmful to bees; indeed its smell seems to repel them; but BHC is lethal and may kill all bee visitors to a crop thus sprayed. Aircraft spraying is the worst of all. You have not much control over the type of spray used by a farmer but it is worthwhile to make friends with him (a jar of honey would be well spent) and get him to advise you of the day on which he proposes to spray. Then you can shut up your bees on the evening before; 24 hours confinement will do them no harm unless the temperature rises above 70° F. Remember also that fruit blossom is not the only source of trouble; rape and field beans are eagerly visited by bees. Both may be treated with insecticides.

ENEMIES

The Wax Moth. The greater and lesser wax moth hover around hives during the summer evenings, particularly in July, and, if they succeed in entering the hive, they lay their eggs on the combs; a favourite place is the saw cut in the top bar of the frame and for this reason wedge-top frames which have no saw cut are to be preferred. The wax moth larvae burrow their way through combs leaving behind them cocoon lined tunnels containing black faeces. If unchecked they will soon destroy combs completely. They prefer old brown combs and live partly on the cocoons left by bee larvae.

The most important defence against the wax moth is strong colonies; then the wax moth has great difficulty in entering the hive and, when it does, eggs and larvae are ejected by house cleaning bees.

Supers may often be stored away with wax moth eggs in the frames and these may hatch out and the grubs work undisturbed during the winter storage. Examine your supers once or twice during the winter and, if the trouble is found, fumigate them. This may be done by burning 1 or 2 oz. of sulphur in an empty brood chamber over which are stacked the supers with their combs. The vapour of tetrachlorethane, used for fumigating greenhouses, is effective.

Many beekeepers place a few tablets of PDB (para-dichlor-benzene) in among their supers during winter storage as a preventive measure.

Wasps. During August and September wasps are very troublesome, entering hives, tearing down combs and stealing the honey. They are stronger and more agile than the bees and can thus often get past the entrance guards; once inside the hive they do not appear to be further molested.

The entrances of the hives must be closed down during these months, if necessary to one bee space, by means of perforated zinc. There must be no cracks or holes in the hive to form an alternative entrance for the intruders. No honey or syrup must be spilt around the apiary. When the first autumn frosts come this puts an end to the wasp menace.

The Bee Louse. A bright red little parasite which lives on the body of the bees, particularly the queen. It seems to do no harm but may be removed by puffing tobacco smoke over the queen when the lice become asphyxiated and drop off. The worst feature of this pest is the fact that its larvae live in the cappings of honey comb and spoil the appearance.

Ants. These insects are often to be found in hives and appear to be tolerated by the bees. If their numbers become excessive a W.B.C. hive may be isolated by standing its legs in tins containing creosote or motor sump drainings; a special stand with legs would have to be made for National or Smith hives.

Mice. To those who live in the country or near open ground, field mice can be very troublesome. In the autumn they enter the hive in the quiet of the evening and make a nest by biting away part of the lower combs, filling the resulting space with grass and dead leaves. Once established they help themselves to the bees' stores and probably only leave the hive in search of water. Apart from damage to the combs, the smell of mice disturbs the bees, they do not cluster properly, use up their stores at an excessive rate and often die of starvation in the early spring. It is curious that the bees appear to make no effort to sting or drive out the mice.

If you have installed entrance guards in your hives early in August and if the entrance is restricted to $\frac{1}{4}$ inch in height mice will not be able to enter. An entrance guard—often referred to as a 'mouse guard'—must remain in place throughout the winter. The presence of mice will be detected by mouse excreta on the alighting board and dead grasses projecting from the entrance. In such circumstances it is best to close up the entrance of the hive completely with a strip of perforated zinc. During the winter months bees may be confined for a whole month without ill effect but the mice, unable to go out and get water, will die.

Woodpeckers. In some districts, woodpeckers are a great nuisance. They attack hives, especially the finger lifts of National Hives, and make great holes in them. The only remedy appears to be to enclose the hive entirely with fine wire netting.

DISORGANIZATION

Queenlessness. Often quite inexplicably a queen will disappear from a hive during the winter period so that, in spring, a stock may be queenless and will dwindle away and die if the loss is not noticed. Some of the reasons for queenlessness are:

Crushing the queen during careless manipulation. When replacing combs in the hive the greatest care should be exercised to avoid crushing bees. The comb containing the queen should not be kept out of the hive longer than is necessary. Avoid opening the hives too early in the season.

Loss on mating flight. Not much can be done about this except to place the mating hive in a special and easily recognizable place so that the queen will have no difficulty in recognizing her own colony on return. Queens seem to mate more safely from small nuclei than from large colonies.

Old age. Do not let your queens die of old age. Replace them when they are two years old.

Queenlessness may often be detected without opening the hive. If the queen has been lost very recently the bees can be seen running up and down the front of the hive as though looking for her. If the hive has been long queenless, the bees become listless and are unwilling to work and the colony is noticeably weaker in flying bees. If the outside of the hive be struck sharply the bees inside buzz or 'roar'; in a queenless colony this continues for a minute or two but in a queenright colony it dies away in a few seconds. If any of the above signs are noticed, open the hive and look at one or two frames from the centre of the brood chamber. If eggs are absent the colony has been queenless for at least three days; if unsealed larvae are absent, 8 days; if sealed worker brood is absent, three weeks.

The simplest remedy is to give the queenless colony a comb con-

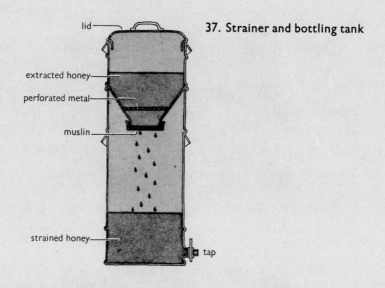

37. Strainer and bottling tank

- lid
- extracted honey
- perforated metal
- muslin
- strained honey
- tap

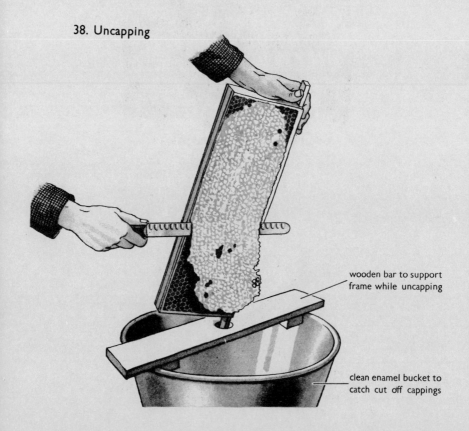

38. Uncapping

- wooden bar to support frame while uncapping
- clean enamel bucket to catch cut off cappings

39. American foul brood (A.F.B.)

40. European foul brood (E.F.B.)

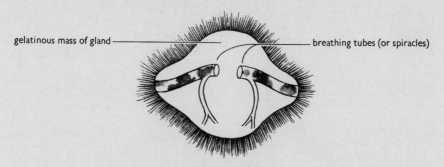

41. Acarine disease

bee after removal of head and first segment of thorax. Breathing tubes are clogged by acarine mites and their refuse. These tubes are perfectly white and clean in a healthy bee

taining eggs taken from another colony; this gives them a chance to raise a queen of their own using one of the eggs. You may purchase a queen and introduce it to the colony. You may unite to the colony a small nucleus containing a laying queen (if you have one). If the colony has been queenless for more than three weeks it is usually better to unite it to a queenright stock and then to start another colony by any of the methods described in Chapter 9.

Laying Workers. When a colony has been queenless for fourteen days or more, the ovaries of many workers develop and they become capable of laying eggs. Since workers cannot mate, such eggs can only produce drones, although the eggs are often to be found in worker cells. The most obvious sign is the presence of typical domed drone cappings on worker cells. Such cells are distributed very irregularly and, if eggs are found, several will often be present in one cell.

The presence of laying workers seems to satisfy the bees; they do not have a feeling of queenlessness and are consequently very difficult to requeen with a live queen. If you have a virgin queen you may allow her to run into the entrance of the hive; if she is accepted and mates the problem is solved but this is a very uncertain method. A better scheme is to put into the hive a frame containing a sealed queen cell, but the best method is to unite the stock with laying workers to a queenright stock, afterwards starting a new colony by any of the methods described in Chapter 9.

Robbing. We have already referred to robbing in earlier parts of this book. When the nectar flow suddenly ceases, out-of-work foragers prowl around looking for further supplies of nectar and they may attack weak colonies and take all their stores. This danger is particularly acute in August and September. When once the robbing instinct is aroused it is not easily suppressed and other and stronger colonies may be attacked. You will notice fighting takes place on the alighting board between the entrance guards and the intruders. Bees can be seen hovering around the hive looking for a chance to slip by the guards or seeking some unguarded crack by which they may obtain ingress. If you attempt to open your hive while it is being robbed (which is a thing you should never do) you will find that the bees are very bad tempered.

Expose no honey or stores which might arouse the robbing instinct; when removing supers, carry out the work at dusk when no bees are flying; when feeding, do not allow any drops of syrup to fall on the ground or run down the sides of the hive. Close up the entrances of your hives early in August by means of perforated zinc, leaving an entrance not greater than 1 inch × $\frac{1}{4}$ inch; if robbing has broken out it may be necessary to reduce this entrance to $\frac{1}{4}$ inch × $\frac{1}{4}$ inch. See that there are no cracks or holes by which the robbers may gain an alternative entrance. Above all, keep your stocks strong.

The Beekeeper. Are you one of the enemies of your bees? Are you a disorganizing force, pulling your hive to pieces too often and without cause? Are you careless in manipulation, crushing and killing bees unnecessarily? Do you take their laboriously gained stores and omit to feed them adequately?

Remember that you are as a god to your bees, that you have in your hands the power of life and death, that you can create conditions of dearth or plenty, that you can work what are to them miracles.

Use that power wisely and with restraint.

Suggestions for further reading

Anon. (1969) *Diseases of Bees*. Ministry of Agriculture Bulletin No. 100.

Betts, A. D. (1951) *The Diseases of Bees*. Bee Books New & Old.

A GLOSSARY FOR BEEKEEPERS

Acarine. A disease of adult bees making them incapable of flight. (*p.* 103).

American Foul Brood (usually referred to as A.F.B.). A disease affecting larvae (*p.* 100).

Artificial Swarming. A manipulation used to prevent swarming, to increase stocks or to raise queens (*p.* 78).

Bee Escape. A one-way exit which prevents bees from re-entering the supers (*p.* 71).

Bee Space. The gap allowed between the frames and the walls of the hive, usually $\frac{3}{16}$ inch (*p.* 46).

Brace Comb. Short irregular pieces of comb joining two normal combs together or filling up gaps (*p.* 28).

Brood. A collective noun applied to the larvae of the bee irrespective of age. If the larvae are not more than five days old they are known as 'unsealed brood' because their cells have not yet been capped over. After this date they are known as 'sealed brood' (*pp.* 28, 29).

Brood Chamber. That part of the hive used for the raising of brood (*p.* 28).

Carbolic Cloth. A piece of material impregnated with a carbolic disinfectant and used to subdue bees (*p.* 54).

Cast. The second or subsequent swarm to issue from a hive. Known in America as an 'after-swarm' (*p.* 38).

Cleansing Flight. An early flight of the bees for the purpose of emptying their bowels of the waste products accumulated during the winter (*p.* 34).

Clear Honey. Liquid honey as first extracted from the cells (*p.* 17).

Clearer Board. A piece of apparatus used to clear the bees out of the supers (*p.* 70).

Clustering. The bees' habit of clinging together in a tight mass. Applied to swarms (*p.* 37) and to the winter cluster (*p.* 33).

Cold Way. A method of arranging frames so that the top bars are at right angles to the entrance to the hive (*p.* 60). See also 'Warm Way'.

Colony. A bee community, including the combs they occupy, brood and stores. The term is usually restricted to a community occupying five or more frames (*p.* 23). See also 'Nucleus'.

Crown Board. A wooden cover placed over the brood chamber or super to prevent the escape of bees (*p.* 52).
Demaree. A system of swarm control named after its inventor. Often used as a verb, e.g. 'to demaree a colony' (*p.* 80).
Division Board. A moveable partition used to reduce the size of the brood chamber (*p.* 85).
Double Brood Chamber. An arrangement of the hive in which the queen is given an enlarged brood chamber consisting of two boxes of combs (*p.* 65).
Drone Laying Queen. An old, imperfectly mated or unmated queen whose eggs, on hatching, produce only drones (*p.* 25).
European Foul Brood (usually referred to as E.F.B.). A disease affecting larvae (*p.* 102).
Extractor. A piece of apparatus for removing honey from cells (*p.* 92).
Feed Hole. An aperture in the crown board or quilt through which the bees are fed (*p.* 53).
Foragers. Bees whose task it is to gather nectar, water and pollen. Usually the older bees (*p.* 26).
Foundation. A thin sheet of beeswax embossed with the pattern of the base of the cells (*p.* 47).
Frame. The wooden rectangular structure within which the foundation is mounted and in which the bees ultimately build their combs. 'A frame of foundation' is a frame containing foundation upon which the bees have not yet built cells. When a beekeeper refers to 'combs' this is usually understood to include the frame (*pp.* 46, 47).
Frosting. A white appearance of the honey largely due to slight shrinkage during granulation (*p.* 96).
Frow Mixture. A remedy for acarine disease named after its discoverer (*p.* 103).
Fructose. One of two sugars present in honey (*p.* 16).
Fully Ripened. A term used to describe honey after the bees have completed their processing (*p.* 17).
Glucose. One of the two sugars present in honey (*p.* 16).
Granulated. A term used to describe honey in which the glucose has crystallized (*p.* 17).
Hoffman Frame. A type of frame in which the widened end members provide the necessary spacing (*p.* 49).

A GLOSSARY FOR BEEKEEPERS 113

Honeydew. An inferior type of honey produced from extra-floral secretions of plants (*p.* 19).

Honey Flow. An old and inaccurate term meaning 'nectar flow' *q.v.* (*p.* 12).

Honey Sac. The organ in which the bee carries back nectar to the hive (*p.* 15).

K-Wing. A sign of acarine diseases (*p.* 103).

Larva (pl. **Larvae**). Young bee-grubs (*p.* 28).

Laying Workers. Worker bees which, in the absence of the queen, become capable of laying a few eggs (*p.* 109).

Lift. One of the outside removable sections of a double-walled hive (*p.* 50).

Manipulations. Operations involving the removal or exchange of the combs of a colony (*p.* 82).

Metal End. One of the many devices used to maintain the appropriate spacing of combs within the hive (*p.* 49).

National Hive. A single-walled type of hive, often referred to simply as a 'National' (*p.* 50).

Nectar. The sweet liquid secreted by certain flowers and eagerly gathered by the bees (*p.* 11).

Nectar Flow. The period when plants yield nectar abundantly (*p.* 12).

Nosema. A disease of adult bees (*p.* 104).

Nucleus (pl. **Nuclei**). A small colony consisting of up to four combs (*p.* 23).

Nurse Bees. Young bees whose duty it is to feed the larvae (*p.* 26).

Out Apiary. An apiary situated away from the home of the beekeeper (*p.* 58).

Parent Colony. The colony from which a swarm has emerged (*p.* 37).

Piping. A shrill high intermittent note emitted by a virgin queen soon after she has emerged from her cell (*p.* 38).

Pollen Basket. A spiny structure on the hind legs of worker bees in which they carry pollen back to the hive. Absent in queens and drones (*p.* 26).

Propolis. A sticky substance gathered by the bees from exudations of certain trees and used to stop up cracks in the hive. Known in America as 'bee glue' (*p.* 31).

Prime Swarm. The first swarm to emerge from a hive (*p.* 38).

Queen Cage. A container in which queens may be sent through the post. The same container also serves to introduce a new queen to a queenless colony (*p.* 88).

Queen Excluder. A grid of wires or perforated metal which prevents the queen from entering the supers while admitting the workers (*p.* 53).

Quilt. A fabric cover placed over the brood chamber or supers to prevent the escape of bees (*p.* 52).

Ripe. See 'Fully Ripened'.

Royal Jelly. A special food secreted by nurse bees for the feeding of queen larvae (*p.* 36).

Sealed Brood. Brood more than five days old sealed over by the workers (*p.* 29).

Skep. An upturned rush basket once widely used to house bees (*p.* 60).

Smith Hive. A simple, modern, single-walled hive (*p.* 51).

Stock. A colony of bees together with the hive in which they live (*p.* 23).

Sucrose. The chemical name for cane or beet sugar. The principal sugar occurring in nectar (*p.* 16).

Super. The upper parts of a hive in which the bees store honey (*p.* 28).

Supersedure. A process by which the bees replace an old queen by a young one without swarming (*p.* 77).

Travelling Box. A small, light, well-ventilated hive used for transporting bees (*p.* 59).

Uncapping. The process of removing the cappings from sealed cells so that the honey may be extracted (*p.* 91).

Unripe. A term used to describe honey before the bees have completed their processes (*p.* 17).

Unsealed Brood. See 'Brood'.

Virgin Queen. A young unmated queen (*p.* 24).

Warm Way. A method of arranging frames so that the top bars are parallel to the entrance of the hive (*p.* 60). See also 'Cold Way'.

W.B.C. A double-walled hive so called after the initials of its designer (*p.* 50).

INDEX

Reference should also be made to the Glossary on page 111. Illustrations have not been indexed but a full list will be found on page 7

acarine, 103
addled brood, 103
American foul brood, 100
ants, 107
apparatus, 56
apple blossom, 63
arabis, 21
aubrieta, 21

bee gardens, 20
  louse, 107
Beekeepers' Associations, 55
bees, drones, 23, 25, 38
  massacre of, 38
  nurse, 26
  queen, 23, 24
    absence of, 108
    drone-laying, 25
    excluder, 53
    finding the, 89
    introduction of, 87
      conditions for, 87
      newspaper method of, 88
      purchased, 88
      Simmin's method of, 88
    mating of, 24
    purchasing, 58
    raising of, 85
    removal of the old, 87
    virgin, 24
  workers, 23, 26
    laying, 109
    uniting, 82
blackberry blossom, 15
blossom, various, 12, 62
borage, 21

bottling, 95
brood chamber, 28
  size of, 35
  sealed, 29
  unsealed, 29

carbolic cloth, 54
cast, 38
catmint, 21
cells, 27
  breeding, 29
  drone, 27
  queen, 28
  transition, 27
  worker, 27
chalk brood, 103
charlock, 13
honey, 19
cherry blossom, 13
chilled brood, 102
cleansing flights, 34
clothing, protective, 43
clover, honey, 19
  red, 13
  white, 13
clustering, 33
  break up of, 35
cold way, 60
colony, 23, 58
  purchasing, 58
  strength of, 35
combs, 27, 46
  brace, 28
  selection of, 93
Cotoneaster horizontalis, 21
cost, 44

crocus, 21, 63
crown boards, 52

dandelion, 13
Demaree system of swarm control, 80
diseases, 100
  adult bee, 103
drones, *see* bees

eggs, 28
  drone, 25
egg-laying, cessation of, 38
enemies, 106
equipment, 56
European foul brood, 102
exhibiting, 98
  appearance of honey for, 99
  clarity, 99
  cleanliness, 99
  colour, 99
  density, 99
  flavour, 99
extractors, 92
  care of, 97

feeding, 18, 66, 68
flights, cleansing, 34
  mating, 24
food stores, 30
frames, 46
  foundation for, 47
  spacing of, 49
  types of, 47
frosting of honey, 96
fructose, 16
fruit blossom, 13
Frow mixture, 103

gloves, 43
gooseberry, 21
glucose, 16
glue, bee, *see* propolis
granulation of honey, 96

harvesting, *see* honey
hawthorn, blossom, 14, 64
hawthorn, honey, 19
heather (ling), 15
  honey, 19
hive, 50
  accessories for, 51
  floors for, 52
  inner covers for, 52
  National, 50
  opening up the, 66
  robbing of, 109
  Smith, 51
  stands for, 52
  tool, 53, 67
  W.B.C., 50
honey, 15, 30
  charlock, 19
  clear, 17
  clover, 19
  containers, 93
  dew, 19
  extraction of, 94
  -flow, 12
  granulated, 17
  harvesting of, 91
    bottling tank for, 92
    cappings tray for, 92
    extractor for, 92
    equipment for, 92
    strainer for, 92
    uncapping knife for, 92
  hawthorn, 19
  heather, 19
  labelling of, 96
  lime, 19
  ripe, 17
  sac, 15
  special, 18
  straining of, 95
  tree, 19
  uncapping of, 93
  undesirable, 20

INDEX

insecticides, poisoning by, 105
ivy, 21

June gap, 15

larvae, 29
lavender, 21
lime, blossom, 13, 65
  honey, 19
limnanthes, 21

management, 62
manipulations, 82
mating flights, 24
metal end, 49
methyl salicylate, 104
mice, 107
mignonette, 21
montbretia, 65

nectar, 15, 30
  first supplies of, 35
  flow, 12
  plants yielding, 12
  secretion of, 11
nosema, 104
nucleus, 23, 59
nurse bees, *see* bees

overcrowding, 73

Pagden method of dealing with swarm and parent stock, 76
paralysis, 105
pear blossom, 13
pests, 106
phacelia, 21
plum blossom, 63
poisoning by insecticides, 105
pollen, 30
privet honey, 20
propolis, 31

queen bee, *see* bees
  excluder, *see* bees

quilts, 52

raspberry, 21
Rhododendron ponticum honey, 20
robbing, 65, 66, 109
royal jelly, 36

sac brood, 103
scent, 31
situation, choice of, 57
skep, 60
smoker, 53
smoking, 66
snowberry, 21
stings, 41
stock, 23
stocks, increasing of, 83
  purchase of, 58
stone brood, 103
storage space, 36
sucrose, 16
sugar, 11, 16
supers, 28, 70
swarming, 36
  avoidance of, 73
  control of, 79
  symptoms, of, 77
swarms, 38, 60, 73
  hiving, 61, 75
  prime, 38
  purchase of, 60
  taking, 74
syrup, preparation of, 68

tank, bottling, 92
  care of, 97
tray, cappings, 92
thyme, 21
thymol, 69

uncapping, 93

veils, 43
veronica, 21

wallflower, 21
warm way, 60
wasps, 107
water, 34
wax, 97

wax, moth, 106
willow herb, 15, 65
winter, preparation for, 39, 66
woodpeckers, 108
workers, *see* bees